The Strategic Management of Technology

CHANDOS
INFORMATION PROFESSIONAL SERIES

Series Editor: Ruth Rikowski
(email: rikowski@tiscali.co.uk)

Chandos' new series of books are aimed at the busy information professional. They have been specially commissioned to provide the reader with an authoritative view of current thinking. They are designed to provide easy-to-read and (most importantly) practical coverage of topics that are of interest to librarians and other information professionals. If you would like a full listing of current and forthcoming titles, please visit our web site **www.library-chandospublishing.com** or contact Hannah Grace-Williams on email info@chandospublishing.com or telephone number +44 (0) 1865 884447.

New authors: we are always pleased to receive ideas for new titles; if you would like to write a book for Chandos, please contact Dr Glyn Jones on email gjones@chandospublishing.com or telephone number +44 (0) 1865 884447.

Bulk orders: some organisations buy a number of copies of our books. If you are interested in doing this, we would be pleased to discuss a discount. Please contact Hannah Grace-Williams on email info@chandospublishing.com or telephone number +44 (0) 1865 884447.

The Strategic Management of Technology

A guide for library and information services

DAVID BAKER

Chandos Publishing
Oxford · England New Hampshire · USA

Chandos Publishing (Oxford) Limited
Chandos House
5 & 6 Steadys Lane
Stanton Harcourt
Oxford OX29 5RL
UK
Tel: +44 (0) 1865 884447 Fax: +44 (0) 1865 884448
Email: info@chandospublishing.com
www.library-chandospublishing.com

Chandos Publishing USA
3 Front Street, Suite 331
PO Box 338
Rollinsford, NH 03869
USA
Tel: 603 749 9171 Fax: 603 749 6155
Email: BizBks@aol.com

First published in Great Britain in 2004

ISBN:
1 84334 041 0 (paperback)
1 84334 042 9 (hardback)

British Library Cataloguing-in-Publication Data.
A catalogue record for this book is available from the British Library.

Typeset by Concerto, Leighton Buzzard, Bedfordshire, UK (01525 378757)
Printed in the UK and USA

Contents

Preface

Technology is pervasive in our world; major technological advances have shaped and changed society out of all recognition. Take the typewriter. In the 1880s, there were only 7,000 clerks (mostly women) in England and Wales. The introduction of the typewriter meant that by 1910 there were nearly 150,000 (Williams, 1982). And the rest, as they say, is history. Since the beginning of the twentieth century, technology in general – and information and communications technology (ICT) in particular – has transformed the way we live. Unless you are a hermit, anyone in western society will use a whole range of technologies – e-mail, World Wide Web, mobile phone, television, satellite, etc. Just look around your place of work and count up the number of 'gadgets' that are indispensable to you.

But it is not just the pervasiveness of ICT that challenges us personally and professionally; it is also the speed at which the technology is changing and developing. You only have to reflect on the extent to which key technologies of today were not in use even five years ago. The Internet is the obvious example. We may think that we have been using it forever, but when did most of us actually start using it? And how many of our children have known any different?

Technology needs managing if it is not to manage us – look at the number of major projects that go spectacularly wrong, such as TAURUS (Drummond, 1999), air traffic control, passport office, tilting trains. Billions of pounds wasted or mismanaged (Collins and Bicknell, 1998). Where new technology has been badly introduced, was it the implementation or the overarching approach that went wrong? There is much evidence to suggest that many of the technology failures that we read about in the papers were as much about inadequate senior management as about operational ineffectiveness and inefficiency.

This book emphasises the importance of taking an effective strategic approach to the management of technology. It argues that operating in this way will minimise the chances of failure, as exemplified by the cases cited above. Developing and implementing a technology strategy, if done thoroughly and thoughtfully, should mean that the many critical success factors that need to be taken into account when managing, implementing and developing technology are recognised as just that and are responded to accordingly.

But effective strategic management is more than that. It is about positioning the organisation to best advantage, taking account not only of its relative strengths and weaknesses and the opportunities and threats that may present themselves over the time period covered by the strategy, but also of its environment and its capacity to invent, innovate, and improve through technology. It also stresses the importance of integration – of strategy, systems, technology, planning, process, people, resources and functions. Above all, it proposes that organisations – and their senior managers in particular – need to 'think outside the box' in order to ensure that they are preparing for the future rather than living in the past.

There are many references in the book to private sector organisations and their approach. Wherever possible, these references are related across to the public sector, where the majority of library and information services (LIS) are located. This proved a relatively straightforward task, given the increasingly competitive and commercial nature of the sector. It is hoped that in referring to the private as well as the public sectors the reader will have a broader set of experiences and tools on which to draw when considering the strategic management of technology. For the public sector manager in particular, it is envisaged that the key perceptions, drivers and approaches to technology management and development will provide a valuable insight into the workings of the technology suppliers with which LIS managers increasingly have to deal. Strategic technology management, like any other form or branch of management, is ultimately about doing. To this end, a number of case studies and practical examples, based primarily on my own experience, have been included.

Bibliography

Collins, T. and Bicknell, D. (1998) *Crash: Learning from the World's Worst Computer Disasters*. London: Simon & Schuster.

Drummond, H. (1999) 'Are we any closer to the end? Escalation and the case of Taurus', *International Journal of Project Management*, 17(1), 11–16.

Williams, T.I. (1982) *A Short History of Twentieth-Century Technology, c.1900–c.1950*. Oxford: Oxford University Press, p. 292.

Acknowledgements

This book grew out of study for a Master of Business Administration degree with the Open University. The qualification centred upon technology management in general and the strategic implications in particular. The work ended with a dissertation on the subject of why technology projects fail. Projects in the LIS environment were studied. Some of the conclusions from the research have made their way into this book. I am grateful to all who supported me during the MBA programme and especially to those who participated in the research project. I must thank MCB Press for permission to reprint Case study 1, which first appeared in their journal *Interlending and Document Supply*, vol. 31(2). I am also grateful to my past and present employers for their support during the time when this book was being researched and written and the earlier MBA research was being undertaken.

List of abbreviations

ANSI	American National Standards Institute
BIDS	Bath Information and Data Service
BPR	business process re-engineering
CSF	critical success factor
CTI	Computers in Teaching Initiative
CURL	Consortium of University Research Libraries
Docdel	document delivery
e-Lib	Electronic Libraries [Programme]
EDDIS	Electronic Document Delivery: the Integrated Solution
FDI	Fretwell-Downing Informatics
FMEA	failure mode and effects analysis
HE	higher education
HEFCE	Higher Education Funding Council for England
HEI	higher education institution
HSA	hard systems approach
ICT	information and communications technology
ILDRMS	Inter-Lending and Document Request Management System
ILL	inter-library lending
IPR	intellectual property right
ISD	Information Services Directorate
ISO	International Standards Organisation
IT	information technology
ITATL	IT-assisted teaching and learning
JANET	Joint Academic Network

JISC	Joint Information Systems Committee [of the UK Higher Education Funding Councils]
JIT	just-in-time
LIS	library and information services
LR	likelihood ranking
MAN	metropolitan area network
MLE	managed learning environment
PEN	plan effectiveness number
PRINCE	Projects in Controlled Environments
PRN	priority risk factor
RAE	research assessment exercise
R&D	research and development
RAN	regional area network
RDA	Regional Development Agency
RLSG	Research Libraries Support Group
SCAITS	Staff Computing and IT Skills
SMEs	small- to medium-sized enterprises
SNEL	Sudanese National Electronic Library
SR	severity ranking
SSM	soft systems methodology
SWOT	strengths, weaknesses, opportunities, threats
TA	technology assessment
TLTP	Teaching and Learning Technology Programme
TQM	total quality management
UEA	University of East Anglia
UKHE	United Kingdom higher education
VCA	value chain analysis
VLE	virtual learning environment
VRE	virtual research environment

List of figures and tables

Figures

Tables

About the author

David Baker was born in Bradford, West Yorkshire, in 1952. His first love was the church organ, which he began playing from the age of 12. By the time that he was 16, he was an Associate of the Royal College of Organists. He gained his Fellowship the following year. In 1970 he was elected Organ Scholar of Sidney Sussex College, Cambridge, graduating with a first-class honours degree in Music three years later. He took an MMus degree from King's College, London in 1974. He then moved into Library and Information Services, taking a Master of Library Studies degree in 1976 and a PhD in 1988. Both of these degrees were from Loughborough University. After a number of library posts at Nottingham, Leicester and Hull Universities, he became Chief Librarian of the University of East Anglia, Norwich, in 1985. He was promoted to Director of Information Strategy and Services in 1995, and Pro-Vice-Chancellor in 1997. He became Principal of the College of St Mark and St John, Plymouth, in July 2003.

David Baker has published widely in the field of Library and Information Studies, with 12 monographs and some 100 articles to his credit. He has spoken at numerous conferences, led workshops and seminars and has undertaken consultancy work in most countries in the European Union, along with work in Bulgaria, the Czech Republic, Hungary, Slovenia, Ethiopia, Kuwait, Nigeria and the Sudan. In recent years, his particular professional interest has been in the strategic management of technology. He gained an MBA degree from the Open University in this subject area in 2002. He has been an active member of the

Joint Information Systems Committee's Committee on Electronic Information and for four years was Chair of the Management Board of the Arts and Humanities Data Service. He has led a number of large technology-based projects in the LIS sector, in relation to both digital and hybrid library development and content creation for teaching and learning. His other key professional interest and expertise has been in the field of human resources. The interaction of technology and human resource management led to his chairing the UK-wide SCAITS (Staff Computing and IT Skills) project. When not working he enjoys watching cricket, walking, creative writing and music – both as listener and performer.

The author may be contacted via the publishers.

1

Strategic technology management

Introduction

The key theme of this book can be summed up in three words: strategy, management and technology. This triangle of subjects encompasses a whole set of challenges, issues, options, techniques and solutions. This first chapter defines the three terms, describes the concepts that underpin them and discusses their constituent elements.

Strategic technology management is a series of activities that are carried out within a particular set of contexts, each having a bearing on the way in which these activities are carried out. The activities are listed in the rest of this chapter and discussed in more detail during the course of the book. There are both internal and external contexts in which strategic technology management works. Internal contexts include the specific organisation, its aspirations, the culture and the people. External contexts include the environment in which the organisation operates, the stakeholders' expectations and the quality standards that the 'industry sector' uses to benchmark success and failure. The prerequisites for effective strategy development, technology management and its strategic implementation are enumerated.

Strategy

How do we define it?

Strategy was first used on the battlefield. It comes from the word for generalship – the art of war. Developing a strategy enables the leader – military or otherwise – to marshal his or her resources in such a way as to be in a better position to win, whether it be a war, a competition or a successful introduction of a new product or service. A strategy is likely to be produced in relation to some kind of challenge – a hostile agent in the case of military strategy, new technologies and market or user demands in the case of this book.

A strategy is the outcome of some form of planning – 'an organized process for anticipating and acting in the future in order to carry out the [organisation's] mission' (Siess, 2002). Corrall (1994) defines strategic planning as:

> Essentially a process of relating an organisation and its people to their changing environment and the opportunities and threats in the marketplace; it is a process in which purposes, objectives and action programmes are developed, implemented, monitored, evaluated and reviewed ... [It is] particularly concerned with anticipating and responding to environmental factors, taking responsibility for change, and providing unity and direction to a firm's activities. It is a tool for ordering one's perceptions about future environments in which one's decisions might be played out.

Mintzberg (1994) emphasises the fact that planning should be a formalised process:

> Planning is a formalized procedure to produce an articulated result in the form of an integrated system of decisions ... thus,

> decomposition of the process of strategy making into a series of articulated steps, each carried out as specified in sequence.

Following such an approach, it is argued, will result in an integrative approach, discussed later in this chapter and elsewhere in the book. In addition, as Stueart and Moran (1998) point out:

> Planning is both a behaviour and a process; it is the process of moving an organisation from where it is to where it wants to be in a given period of time by setting it on a predetermined course of action and committing its human and physical resources to that goal.

A strategy that does not drive and facilitate change and improvement is of little use. It therefore also requires an implementation process. The ways in which strategy can be implemented are discussed in detail later in this book. However, it is important to stress at this stage that there has to be a close fit between the high-level strategy and the actual day-to-day operations on which an organisation depends for survival. Tensions can arise between strategic and operational management, with the one emphasising the long term, the other the immediate requirement. In practice, operational management should be at least partly concerned with what will happen more than 12 months ahead, and strategic management cannot be solely about what life will look like in five years' time.

The need for an integrative approach to the strategic and the operational is particularly important in the case of technology management, as noted later. The strategy should coordinate the various operations that are carried out, either as ongoing activities or as finite projects, in order to ensure that they are all contributing to the organisation's overarching mission. Above all, a strategy should enable managers to make decisions, within the context of risk analysis and management. A strategy document should aim to provide the route map by which an organisation, its leaders and its people can chart a course through the issues and the environment in which they will be operating over the course of the planning period. As Drucker (quoted in Siess, 2002) points out:

> Strategic planning is the continuous process of making entrepreneurial – or risk-taking – decisions systematically and with the greatest knowledge of their future consequences; systematically organizing the efforts needed to carry out these decisions; and measuring the results of these decisions against the expectations through organised, systematic feedback.

Strategies need to be flexible in order to cope with future uncertainty and a sure knowledge that the only thing that we can accurately predict is that our predictions will be proved wrong. However, although a strategy does not of itself predict the future, it has to be based on such predictions in order to provide the necessary route map. Later chapters of this book will look in detail at how futures can be predicted in a meaningful way so that technological development becomes the tool rather than the dominant force in strategic development within an organisation.

Because strategy is so contingent upon the specifics of the environment in which it is formulated, there is no one generic model that can be identified and applied across different sectors, countries or technologies. It is unlikely that library and information services (LIS) staff will be working in a 'greenfield' site where there is no competition and where the first entrant can dominate the sector, setting the technology standard for others to follow. It is much more the case that in the LIS sector there will be a high level of existing engagement with technology and increasingly high expectations of service providers. There will be a number of constraints in terms of resource availability, political and social pressure, and organisational culture, among others.

Strategic planning will also need to take account of the organisation's capabilities, as against those of the rest of the sector. Strategy, then, has to be worked out anew for each set of circumstances in which a strategic approach is required. There is nevertheless a wide range of generic techniques that can be applied in different circumstances and a number of these are discussed in later chapters. They encompass the key elements of strategy formulation, planning and implementation and include forecasting, scenario planning and risk management.

Why do we apply it?

Strategy is typically applied so that better decisions can be taken, especially for the longer term.

> By planning, an institution develops both a vision of the future and a strategy for controlling it. This enables the employees to anticipate rather than react to events in the future. In addition, the planning process itself is very important. Properly done, the employees in the institution develop and agree on common goals; that is, they make sure they are on the same wavelength and are moving together in the same direction and toward the same ends. Although the usual end result of strategic planning is a document, the best results are obtained because of the process. (Siess, 2002)

This emphasis on anticipation allows the organisation to respond successfully to threats and opportunities by being better prepared. This preparedness will usually take the form of strategic decision making in order to position the organisation to best effect within the sector, the market and the technology environment.

Crawford (1991) gives four reasons why strategy is so important in an organisational context. It focuses team effort, it brings integration across the organisation, it enables senior managers to delegate (in the knowledge that the strategy gives them and the rest of the workforce a framework within which to operate) and it requires the leadership of the organisation to be proactive rather than reactive.

> If it is necessary to state what a project's focus will be, chances are the investigation of the opportunities will be more thorough. And, if the strategic statement must include all critical guidelines, then its author had better study the selected opportunity thoroughly. In other words, having to write out strategy helps create better managers. (Crawford, 1991)

What are the key elements?

Vision

It is now commonplace for organisations to develop a long-term vision of their future, in order that they can plan their operational and organisational strategies and empower their staff to develop their areas of activity for the good of the business. Any vision statement must be such as to allow for this empowerment. If an organisation has a vision, then it can start to plan the effective deployment of resources, not least against any competition that the organisation will experience within the sector. Major technology innovations are typically the product of someone's vision. Microsoft was founded on one man's vision – a vision of 'where to go' and what the next-generation systems would look like. It need not be one person's vision, however. A whole group of different types of stakeholder might develop and share a vision. This is likely to be true where the vision is born out of the strategic planning techniques described and discussed later in this book.

A high-level vision is seen as a prerequisite of a good strategy. Siess (2002) defines a vision as:

> A concise statement of what [an organisation] would like to become in the future. It should be so idealistic that it is not attainable in the foreseeable future.

This definition throws up a tension between the idealistic and the realistic. A vision that is totally unachievable is hardly worth considering. On the other hand, a vision that is too easy to achieve will neither stretch the organisation nor provide the basis for a strategy that will allow an effective response to future changes, especially when technology trends are likely to prove increasingly demanding. During the 'visioning' process, therefore, it is likely that several options or scenarios will be identified that will require assessment, and one or possibly more preferred longer-term states chosen as the basis of the mission and as the underpinning of any technology choices subsequently made.

Mission

The vision is articulated in the form of a mission statement. Bart (1998) argues that the statement is a formal written document that captures 'an organisation's unique purposes and activities'. Siess (2002) defines the 'mission' as:

> A relatively short, clear statement of the primary purpose(s) of the [organisation]. It consists of the [organisation's] reason for being, what it does, how it does it, and how [it] is different from its competitors. The mission should reflect the organisation's values or the basic beliefs to which [it] and its stakeholders have agreed. It may be achievable in the medium term.

Mission statements are normally made up of several elements. The *aims* of the organisation form the basis of both the statement and the strategy. Ultimately, a strategy has to support the achievement of the aims that an organisation has set itself in order to realise its vision and serve its mission. Senior managers need to outline strategic goals. They also need to check that their goals are consistent with one another. They need to ensure that there is organisation-wide ownership of their plans. Nobody starts with a blank sheet of paper – so they need to know what their starting point is, how far they go and what to do to turn theoretical goals into practical reality. These aims or goals are not framed in terms of specific technology applications or solutions, but are generic, covering the whole of the strategic planning period. There is much evidence to suggest that many technology applications focus too much on short-term rather than long-term objectives. Visions and missions may be driven by *values*, which are likely to determine particular strategic choices. Adopting a 'green' policy, for example, will have a significant impact on any strategy that an organisation chooses to develop and adopt.

There is often confusion between aims (or goals) and objectives. The Higher Education Funding Council for England (HEFCE) has defined an aim as being 'the high-level strategic outcome towards which [the organisation is] working throughout and beyond the life of [the strategic] plan', whereas an objective is something that the organisation seeks to achieve during the course of the plan period, with an indication of the target and the means by which it will be achieved (HEFCE, 2003). However, the term *objective* is often used synonymously with the word 'aim'. Objectives are typically organised into a hierarchy. *Primary objectives* are broad, visionary, fundamental and overarching. They could also easily be described as aims. They are supported by *strategic objectives*. These are long term, and serve the primary objectives. *Operational objectives*, on the other hand, are short term and serve the strategic objectives. In an academic LIS context, for example, a primary objective might be the delivery of leading-edge services to users; a strategic objective might be to be delivering 80 per cent of journal article content electronically within three years of the start of the strategy; and an operational objective would be to wire up all the students' bedrooms to accept broadband transmissions.

A description of the *ways in which the vision will be achieved* is normally included. These may be described as 'operating statements' or a set of tactics – the group of actions that enable the strategy to be implemented successfully. It is the operationalisation of the strategy and will include shorter-term strategies, policies and plans. At this level, some description of the technology options and possible solutions is likely to be included. This description may also refer to key *policies* that the organisation will need to adopt in order to implement the strategy. These policies will ensure consistency of application of the strategy, as for example with equality of opportunity or disabled access. There might also be a *commentary* that gives more detailed explanations of the aims and objectives of the organisation and the principal activities through which progress will be made. This might incorporate the key elements of a *technology roadmap*, outlining where the organisation is currently on the developmental spectrum and where it needs to be.

The description should include *key performance targets*, by which the organisation can demonstrate, in measurable terms, progress towards the stated aims and objectives, together with the *milestones* by which they should be achieved and the *measures* by which performance will be evaluated against target. These should be complemented by a list and description of *key risks and dependencies* that the plan takes into account and that are specific to the key aims, with an outline of how these are to be managed. Risks and dependencies should also include a note of *constraints* that are likely to limit the strategy, its key aims and their delivery.

A *position statement*, describing the organisation's current state, its services or products, its relationship to its competitors and, perhaps, its level of technology usage and awareness is also included. Such a statement might also usefully include reference to the organisation's key *stakeholders* – those who have a particular interest in the health and future development of the organisation. Stakeholders will include staff and customers/users, but may also encompass governing bodies/shareholders, government agencies and professional bodies. Lists of stakeholders may also need to refer to other organisations – whether partners or competitors – working in the same sector. The relative importance of different objectives may be determined by the priorities of the stakeholders in general and by those of the dominant stakeholder or stakeholders (discussed later in this book) in particular.

Resource allocation

Strategy and resource allocation go hand in hand. A strategy without a budget is incapable of being implemented. Resource allocation is an operational issue, and one that is typically addressed as part of programme and project management, discussed later in this book. Where major costs are involved, the balance of spend between 'the centre' and departments may pose problems when attempting to institute organisational change. Indeed, particularly where institutions are highly devolved with departmental cost centres, institutional policies derived from strategic institutional management will generally need to be championed by senior managers who will work hard to 'sell' the advantages to the various sections of the organisation. Even if the departments are not required to pay from their own budgets, they pay indirectly via a 'top slice'.

Benefit and risk

Any strategy must take account of the benefits and disbenefits of implementation and of the risks associated with undertaking the work of carrying out a strategy or, indeed, of not doing so. This is especially true of technology projects, where failure is often more likely than success and risk has to be especially well managed. Once a decision has been made to proceed, on the grounds that the advantages of proceeding outweigh the disadvantages, then effective and efficient project management is a critical factor in success. Risk and project management are significant subjects, considered in later chapters.

What is the value of mission statements?

Mission statements do not in themselves provide the strategy or the solutions for the institution (e.g. they do not describe in detail how technology will be used to best advantage). They can make the thoughts of senior staff of the organisation sufficiently explicit for other staff working in the business to understand what the organisation is trying to achieve. To a certain extent, they can also explain why particular strategies are adopted and what groups' and individuals' particular contributions to the vision and the goals are meant to be. In addition, the statement can be used as the basis for much of an organisation's external communication (Klemm et al., 1991). Ideally, they provide the fundamental tool by which an organisation sums up its basic aims, objectives and core values for all its stakeholders. The statement is also useful as a means of turning the vision and the aims of the organisation into reality.

Too many organisations pay lip service to the idea of mission statements, engaging in a process that produces a 'lowest common denominator' approach to strategic management of the organisation that leaves most personnel just as confused as if there were no mission statement. It is too easy to mistake acquiescence with the wishes of the various stakeholders with the need for ownership of the mission and the vision. The result is neither vision nor ownership, but a vague generality that cannot be used to gear up the organisation for the future; worse than that, the result can be greeted with cynicism by the 'shop floor' in particular and make subsequent attempts to introduce real strategic management more difficult.

The good mission statement is difficult to write, not least because it should say what the organisation is not going to do as well as the areas on which it intends to focus. The longer the statement is, the less likely it will be that it has been thought through carefully and clearly. Writing the pithy, encapsulating statement therefore requires a sufficient investment of time and energy – and a willingness on the part of senior and middle managers – to engage with the formulation process. There will be a trade-off between the amount of time spent and the usefulness of the statement to the organisation, not least in high-risk areas or volatile sectors where rapid change and the need for flexibility may make the application of set objectives beneath a mission statement inappropriate. There might also be a tendency in such circumstances to produce such general statements (because they do not commit the organisation to a particular pathway) that they become 'motherhood' statements, meaningless to the workforce. There is a large subjective and cultural element in the area of mission statements. However, as Robert (1993) implies, one public test of the effectiveness of the statement is the extent to which it could be used to advertise the product or service.

What are the prerequisites for effective strategy development and implementation?

It is important to ensure that the organisation is able to implement the strategy, once devised and agreed. Many strategies fail because there is too much of a gap between conception and reality. Before a strategy is developed, therefore, it is important to see if the organisation is ready to meet the strategic challenges required of it. 'Because [strategic] planning is a delicate, complicated, time-consuming process, it cannot be forced on an organisation that is not prepared for self-analysis and the change that will result from the process' (Stueart and Moran, 1998). Tregoe and Zimmermann (quoted in *The Pocket MBA*, 1992) suggest a 'strategic IQ test' to see if the organisation is 'fit for purpose'. Whether or not this approach is taken, it is important for senior managers to check the following:

- The clarity of their longer-term vision and the appropriateness of the associated strategy. (Do they understand what they want to achieve and how they are going to achieve it?)
- The extent to which the organisation *as a whole* is clear about the vision. (Do staff know what the organisation is trying to achieve?)
- The extent to which the organisation *as a whole* is aware of the key elements of the strategy and understands the reasons for it. (Can you ask any member of staff about where the organisation is going and obtain a coherent answer that fits with the senior management view? Do employees know the contents of the strategy and understand why it says what it says?)
- The strategy is based on extensive analysis of the environment and key future trends. (What are the key factors affecting the organisation currently and how are they likely to change or stay the same in the period covered by the strategy?)
- The strategy drives the decision-making and resource-allocation processes. (Are decisions taken in such a way that they complement rather than contradict the strategy? Are money and people distributed in line with the key objectives of the strategy?)
- The extent to which all parts of the strategy are joined-up with each other. (Do the various aspects of the strategy make sense when put together? Are they well integrated, as for example

staff training and development programmes in relation to the implementation of new technology?)

- The strategy's progress is regularly monitored and the results used to change direction or improve performance towards the goals set. (Does the organisation regularly ask itself: 'How are we doing in relation to what we said we would do?' 'Do we need to do better or differently?' 'Are we still going in the right direction?' 'Have things changed to the point where we need to change the vision and/or the strategy?')

Maintaining or developing an organisation in which strategic thinking can flourish is a critical success factor in technology management – and, indeed, management more generally. So too is the visionary ability and capacity of the organisation's leadership in particular. In recent years, the need to adapt to rapidly changing circumstances and to marry operational with strategic need and thinking has led to the formulation of the concepts of organisational development and the learning organisation.

Organisations that learn from their mistakes are much more likely to be successful in their development and implementation of strategy that those that do not. 'Lessons learned' reports are useful at all levels of an organisation in this respect. In terms of strategic planning, it is crucial that the underpinning techniques used allow the organisation to set an appropriate general direction, anticipating opportunities and threats, using all relevant information – information gathered effectively and efficiently. This direction has to be communicated well across the organisation and to all its stakeholders and markets. Above all, however, the organisation's leadership has to 'make things happen' and be prepared to change direction when analyses of the current strategic position require firm action of this kind.

The literature of technology management talks about strategic frames and strategic formulae. Frames are a set of beliefs about the organisation's external environment and the organisation's competitive position within it. Formulae are the perceived best practices that will enable the organisation to develop and implement strategy effectively. The strategic frame will be formed from the background, training, experience and values of those who take the decisions within the organisation. Both frames and formulae may well encompass groups of organisations or whole sectors. This is especially likely to be the case in the public sector and LIS units, where the overarching frames of reference are likely to be set by funding agencies – including the government itself – or those bodies that commission technology development, particularly in pre-competitive situations. Senior strategy managers within organisations may well serve on national or regional committees charged with developing public policy in these areas and their local and national frames and formulae will in consequence overlap to a high degree.

In order to plan ahead, it is also important to know about previous activity in the area and to analyse the key factors that are likely to influence the success or failure of the new strategy. The implementation of a strategy takes place within a given environment. Without knowing that environment sufficiently accurately – both now and, as far as is possible, in the future – it will be impossible to determine a strategy that will be truly effective. And given that the strategy aims to introduce something new or different within the organisation or to its users or customers – whether it be process, product or service – then the characteristics of what is to be offered have to be identified and matched against the environment – the 'market' in which the 'product' is to be 'launched'.

Finally, it is important to stress that the implementation of a strategy has to be supported by resources. These might be financial, but they are also likely to include time and commitment, especially of the senior staff who are ultimately responsible for developing and implementing the strategy. The appropriateness of the level of resource to be made available will be a particular issue in major innovation projects, when research and development may well require a particularly intensive and extensive investment.

Technology

How do we define it?

The root of the word 'technology' is given as the Greek word 'tekhne', while 'logy' comes from the Greek 'logia/logos'. The Greek 'tekhnologia' is perhaps best translated as 'systematic treatment'. Technology is not just an academic discipline or a science in its own right. It is also a series of systems, processes or methods that have been developed to undertake a given set of activities more efficiently or more effectively.

Although we tend to think of technology as a modern concept, it is one that can be traced back to at least the eighteenth century and the beginning of the Industrial Revolution. The 'spinning jenny' was a piece of technology; so were the canals, in that they represented systems designed to solve a particular problem or meet a set objective using artificial methods. In the case studies at the end of this book, I look in more detail at the ways in which technology has been applied to specific areas of library activity.

What are the key aspects of technology?

Firstly, technology is a body of knowledge that is continuously changing and growing. It is an 'ology' like sociology and epidemiology, and a whole range of other subjects that are researched, taught, studied and commercially exploited through the activities of universities and research and development (R&D) divisions of commercial institutions. Technological know-how is a valuable, if intangible, asset that can make the difference between success and failure for many organisations.

Secondly, technology is a method concerned with the understanding, development, implementation and use of systems that aim to solve problems. The basic aim is utility or usefulness. As emphasised later in this book, technological method is concerned with the whole environment and not just the 'hardware' that forms the basis of the system being developed. The interaction between people and systems, for example, is of crucial importance in technological method; so too is the dynamic behaviour of the technology system and in particular the transformation process that is at the heart of it (see Figure 1.1). It is about analysis and design with a view to either improving an existing system or developing a new one. And above all, technological method is about managing change. Changes through technology can take a number of forms. They include anything from minor improvements to existing systems to landmark changes – usually through the introduction and widespread adoption of an invention or an innovation – that completely transform the way in which work is carried out and/or people and institutions behave.

Thirdly, a technology is a system that has been developed to serve a specific purpose. A system consists of components, each of which has a logical and meaningful connection to the others and is integral to that system. The removal of even the smallest 'cog' will change the way in which an integrated system works. At the heart of a technological system is the transformation process.

Figure 1.1 The transformation process

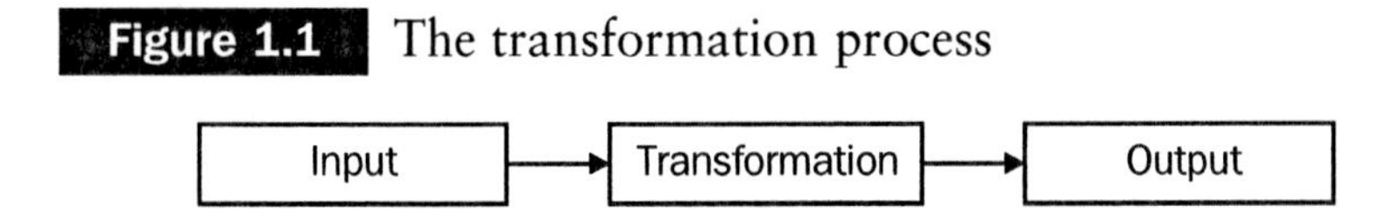

What is the transformation process?

A reader's request for a journal article is transformed into a delivered item through a transformation process. Typically, this process has been associated with industry, where there is a beginning, middle and an end to a series of activities that result in something useful being made. However, the concept is relevant to any technology application. So, the bibliographic citation is the 'raw material' that the 'manufacturing process' turns into the 'finished product' of a document delivered to the end user.

Inputs and outputs can be of two kinds: primary and secondary. A primary output is the product, service or effect that the system was designed and devised to achieve. A primary input is the thing that is transformed to produce the output. Secondary inputs are the 'raw materials' that the transformation process requires in order for primary inputs to be turned into primary outputs. In the case of document delivery processes, it could be computer hardware, library assistant time or a bibliographical database. Secondary outputs are typically what is left when the transformation process is complete, as for example a record of a completed inter-library loan request (see Figure 1.2).

Figure 1.2 Primary and secondary inputs/outputs for document delivery

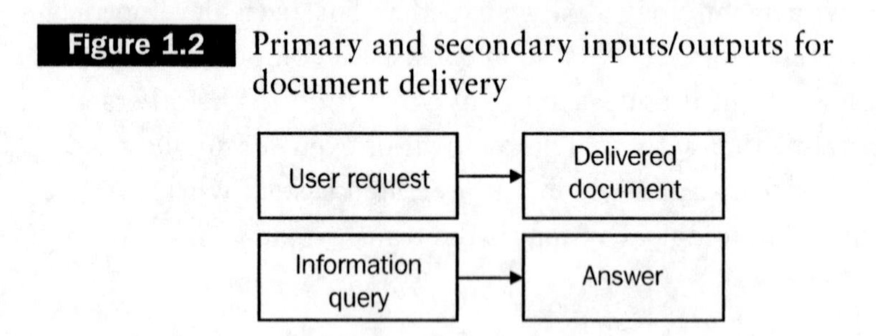

The transformation process is a managed rather than a natural process. It needs to be planned, monitored and controlled as part of a larger system or systems. This control process needs to:

- set the transformation process into a broader context;
- oversee level and quality of both primary and secondary outputs;
- benchmark and compare output levels to set targets;
- take preventive or corrective action when and where appropriate;
- provide a reporting mechanism – typically to senior management.

In our example of document delivery, this work will be undertaken by people rather than by technology. The technology will provide much of the data on which the control process is based, however. So, for example, the reporting to senior management will use statistical information produced from the computer system used to process the document delivery requests.

The basic model of a technological system can therefore be described as in Figure 1.3. This model can be applied to a whole system, or a series of subsystems that make up the system. The principles remain the same.

Figure 1.3 Elements of a technological system

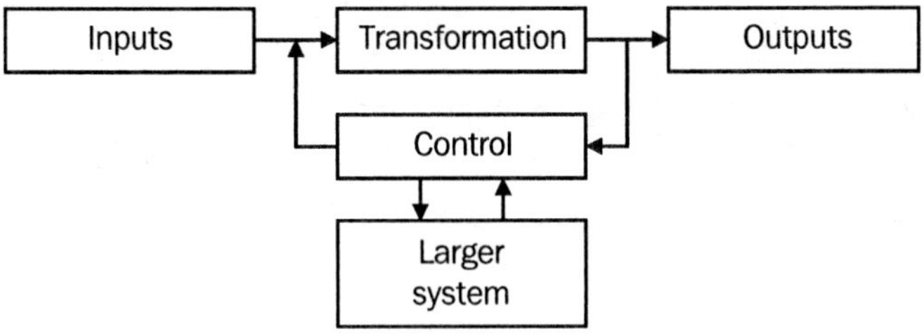

What are technology trends?

Twiss and Goodridge (1989) postulate that the development, exploitation and market adoption of a given technology is not a random event or series of events; instead, it follows a clear pattern that can be forecast, plotted and monitored as the basis of an organisation's or a sector's response over time. This plotting of trends typically takes the form of a standard x,y graph, where y is the performance of the technology in market terms and x is either the cumulative investment or the time over which the technology's development and market penetration is being analysed (see Figure 1.4).

Figure 1.4 A linear technology trend

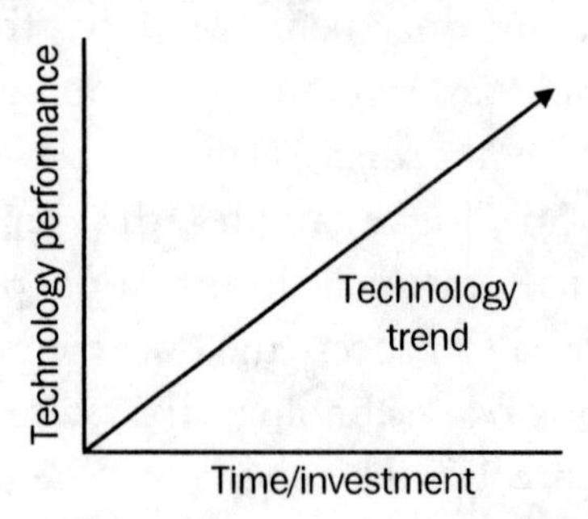

In reality, a technology trend, duly plotted on the graph, is unlikely to be linear. However, there are a few basic trends which have been identified over time and which can be used as a reasonable predictor of the future. The S-curve is one such trend, as plotted on an x,y life-cycle graph (see Figure 1.5)

Figure 1.5 S-curve technology trend

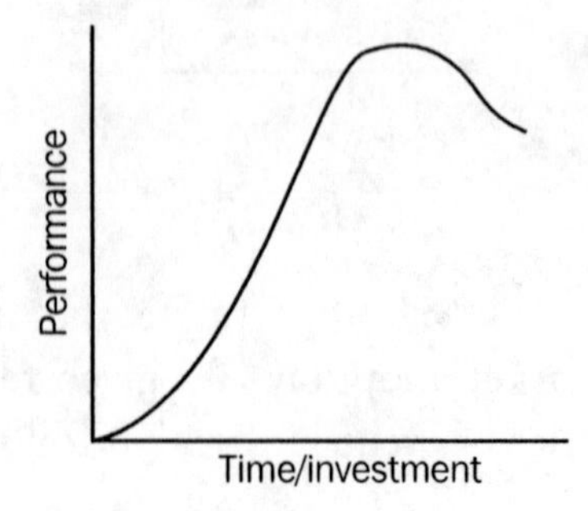

In the early stages, progress of the technology and its market penetration is slow and limited. There may be a low or even negative return on investment of those 'backing' the technology. Where there is success, it may be in narrow or specialised fields, perhaps with support from government or other agencies in order to stimulate innovation and markets and to protect the embryo product from too much competition too early in its life cycle. The flatness of the curve may last for several generations before the technology 'ripens' to the point at which it can 'take off' into mass markets or improved performance or both. In the meantime, the product may not be particularly robust; the emphasis is on future potential and the management of risk associated with failure – of the invention, its innovation into existing or future markets, or of the markets themselves. Some products may never reach the second stage and the second stage is either absent or the curve remains relatively flat. In such cases, any investment has to be written off and as many lessons as possible learnt from the failure.

If the invention/innovation reaches the second stage, then rapid growth occurs. This gives the curve its characteristic 'S' shape. The growth rate is typically characterised by a quick succession of new products that improve the performance of the prototypes and the earlier production models. Product life spans and early obsolescence are prevalent. Competition for superior positioning within the markets will be intense and in commercial areas, companies will aim to cut development times in order to secure and expand their niche. There will be an increasing diversification in the application of the technology and new companies that aim to spread the markets for the products or services will spring up. Gradually, any technical uncertainties will be resolved and a dominant design will emerge. The overall market will continue to grow, but growth rates will slow, and there will be an increasing emphasis on market segmentation and specialised product design to meet the particular needs of the various segments. Within the overall dominant design, product lives grow longer, and the cost –

to both supplier and customer – becomes more important, not least as a means of differentiating one variant of the core product from another.

In the third and final stage, the product or service reaches maturity and, eventually perhaps, saturation. The scope for further innovation is very limited, and any remaining competition is likely to be on price grounds alone. A high investment price has to be paid for any further innovation that is possible, with the emphasis being on reduction of production costs and/or improving quality to maintain the attractiveness of the product or service. There may be price wars, sometimes accompanied or followed by mergers and takeovers. Often, a few dominant suppliers emerge.

What effect do political and regulatory factors have?

Government support or intervention can have a significant impact on the extent to which new technology is actually successful. This intervention can relate to the stimulation or depression of markets, the patenting process, the ability to support embryo technologies and the extent to which competition from other countries or regions can be repelled, at least until the innovation is sufficiently robust to face that competition with a fighting chance of success. Sudden discontinuities can cause significant changes in the overall environment in which technologies are introduced and used. 'Green' factors, for example, have had a major impact on the ways in which organisations in the western world now operate and this has affected users' requirements of technology. Legislative change relating to employment is another area in which technology applications have had to be adapted in order to respond to environmental change.

The impact of political and regulatory factors will vary from sector to sector, industry to industry and country to country. In addition, individual institutions will respond differently to government and related initiatives. The larger an organisation, the more likely it is that it will be affected by political and regulatory considerations. Indeed, in the case of the vast multinational corporation, it may be the other way round, with the corporation affecting government policy and legislation. Think of the influence of a company such as Microsoft, for example.

Given that a majority of LIS units are in the public rather than the private sector, there is the potential for significant involvement and intervention by government. This can have a positive influence on technology development in that governments become commissioning agents, whether directly or indirectly, for R&D work and its subsequent commercialisation, exploitation and application. The degree to which an interventionist approach is healthy will depend upon the nature, robustness and maturity of the technology and its putative markets. Whatever the nature of these dependencies, there will come a point when the technology has to face the markets for which it is designed or where it has the potential to succeed.

What are technology life cycles?

An important aspect of technology and technology management is the life cycle. This is not the same as a life cycle in the accounting definition, in which an attempt is made to take into account all of an asset's costs during the lifetime of that asset (Snyder and Davenport, 1997). A technology life cycle is likely to be much longer term and may include several costing cycles. New technology has to be invented, often after extensive research and development work. It is then developed into a saleable product, typically after much design effort.

The exponential growth in technological know-how and its application means that product life cycles have shortened considerably in the last decade. 'The ubiquitous devices and systems of the turn of the 21st century were all less than 20 years old in 2000, and some of them had less than five years of history behind them' (Feather, 2003). This means that the innovation process itself has had to change, with many developments running in parallel with each other rather than sequentially. The built-in obsolescence that has long been a significant feature of technology development has escalated to the point where product 'life expectancy' has been cut to months rather than years, especially in areas such as fast-moving consumer goods (e.g. television sets).

There is then a variable period of time during which the technology is, or is not, adopted by the market in which it has been launched. If the technology takes a hold, then a broadening range of implementations will occur, including markets and environments for which the technology was not originally intended. The time between the widespread adoption of a technology and the effect that it has on changing working practices, business processes and improving effectiveness and efficiency may be considerable. The initial introduction of computers did not necessarily improve processes or productivity. Only after several years of broadening access to personal computers and network connectivity did automation begin to make a difference.

There will come a peak in the technology life cycle where the impact is at its maximum, after which there will be a reduction in popularity and usage, unless there is a revision of the technology to respond either to changed demands or to alternative products. When the technology starts to become obsolete, a new life cycle will begin. In Figure 1.6, it is assumed that the life cycle begins with an invention each time. This is not necessarily the case. Other forms of 'new' technology development – such as innovation – will be explored in later chapters.

Figure 1.6 A technology life cycle

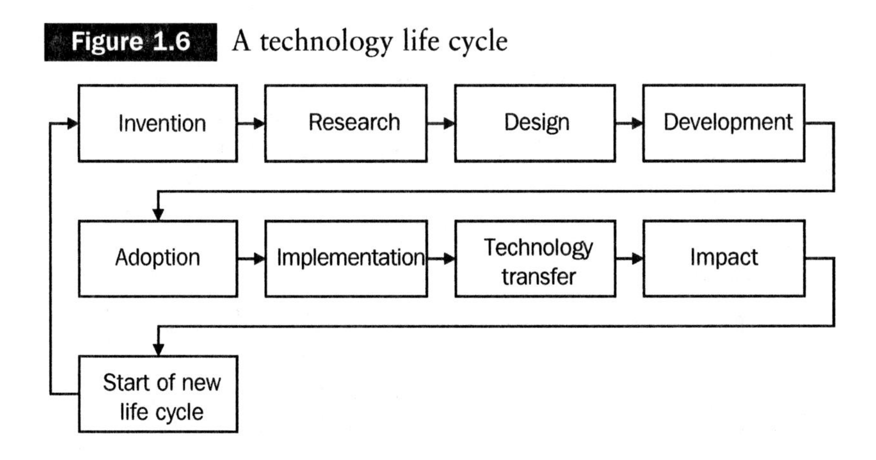

What is technology substitution?

Substitution of design features can extend a technology life cycle. Given that the life cycle is limited by the performance of the technology, a related technology that performs better than the original can extend both the lifetime of the product or the service and maintain and perhaps even expand the core markets. The related technology may incorporate new, higher-performing components into an otherwise mature product or service, as with for example web-based front-ends to older-style library housekeeping systems. Moving from one technology to another in this way can be a traumatic experience for both suppliers and users. There may be a reluctance to change, not least because in the initial stages of the new technology, the existing product or service performs better than the substitute and the costs associated with the latter are higher than with the more robust, older version, as was the case for UK higher education (UKHE) with a possible transfer from ILLOS to EDDIS, as discussed in Case study 3. Proposed substitutions may be met with a degree of cynicism by both suppliers and users, not least because of previous bad experiences with new technology, and especially those that did not live up to their initial promise, such as the EDDIS system. Forecasting using S-curves can help an organisation to decide when to make a technology substitution. This will vary from sector to sector and situation to situation, not least depending on the extent to which the organisation wishes to be a leading-edge player or a fast/slow follower of technology leaders. Timing is everything.

One way of assisting with the decision-making regarding the introduction of a new technology is to trace the extent to which the proposed new product or service is replacing the old within the relevant market sectors. Plotting the substitution rate between old and new can help an organisation to determine the optimum time at which it should make its own technology substitution (see Figure 1.7).

Figure 1.7 Substitution of competing technologies

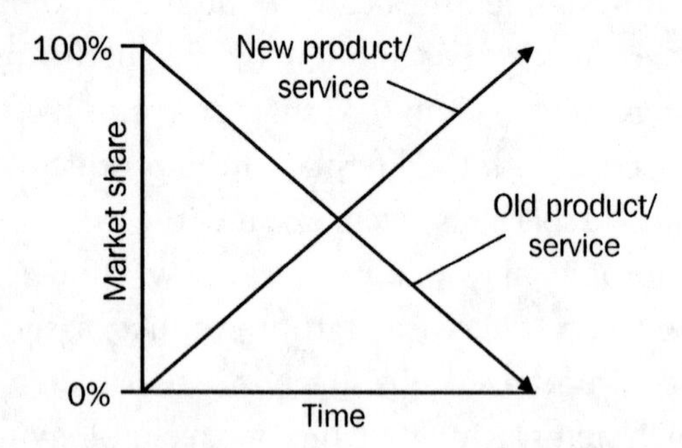

As before, in reality the trend is unlikely to be expressed as a straight line, but rather as two S-curves, with that relating to the old technology/product/service being a reverse S. A time after the cross-over point is arguably the best for a more cautious organisation to make the desired substitution.

Technology forecasting is hampered by the fact that the introduction of an innovative product may well stimulate the suppliers of existing technologies to improve their own products or services to the point where they can compete effectively with the innovation or invention, such that the producers of the latter find it difficult if not impossible to break into the markets that are pervaded by the existing dominant design. However, as Steele (1989) comments: 'If you examine the history of technology, you are forced to conclude that all technologies are fated to be replaced – eventually; however most attempts to replace them will fail.' The question is not when, but where and how. As noted earlier, it is all a matter of timing. This is where good strategic technology management – based on the kind of technology forecasting described here – comes in. Technology assessment (TA) is discussed in more detail in Chapter 3.

What are the different kinds of technology?

Product technology is concerned largely with the primary output of the system. The technology is fundamental to the output: software, hardware, tilting trains, swing-wing aircraft. The technology creates an innovation, typically of something that can be applied to a particular objective or activity – a product.

Process technology is concerned with transformation rather than output. It forms a secondary input to the transformation process. An Inter-Lending and Document Request Management System (ILDRMS) is an example of a process technology.

Control technology is concerned with the automatic manipulation and communication of information.

Technology can also be categorised according to the different ways in which it can be used for strategic competitive advantage, as shown in Table 1.1.

There are other ways of effecting this categorisation. An enabling technology could also be described as a *base technology* as it underpins the whole of the sector but is not critical to the competitive position of any one part of that sector. A critical technology can also be labelled as a *key technology*, because it does affect the relative competitive position of individual organisations within the sector, whereas a strategic technology can also be known as a *pacing technology*: it is at an early point in the

Table 1.1 Technology categories

Technologies	Description
Critical	Those that are central to an organisation's competitive position, which are proprietary to some degree, and which differentiate it from the competition
Enabling	Those that are not proprietary to the same degree, are broadly available to all members of the industry, but are essential to the efficient design, manufacture and delivery of the organisation's product or service, and its level of quality
Strategic	Which can be emerging or already existing technologies, whose salience arises from their ability to provide new competitive opportunities when combined with or substituted for existing critical or enabling technologies

(Adapted from Whelan, 1989, quoted in Open University, 1994.)

S-curve, but it is a technology to be monitored closely in that it may well affect the future of the sector, its markets and its approach to technology application. Clearly, key and pacing technologies are those where forecasting is most crucial, although it should be remembered that they in their turn may supplant existing base or enabling technologies.

How do we acquire technology?

There are various ways in which technology can be acquired. In the LIS field, the likelihood is that technology will be acquired through direct purchase from a third party. In other words, there will be a relationship with a supplier that is external to the organisation. Some form of contractual arrangement will be made. This is a perfectly acceptable approach for existing technology. Technology development and innovation (discussed in Chapter 2), on the other hand, may require a more sophisticated and extensive relationship or framework. Pavitt (1988) talks of organisations whose technology base has developed incrementally over time; these are described as 'investors'. Others acquire their technology base, perhaps by takeover; they are labelled 'traders'.

A contractual relationship may be struck between a commissioning organisation and an R&D unit to undertake specific development. Such has been the case in UK higher education in respect of the work undertaken by the Joint Information Systems Committee (JISC). It could be said that much of the work that has been done by JISC has taken the form of joint ventures between libraries and technology companies. There are many variants of this, including the setting up of consortia, the interchange of personnel and the licensing for further development of pre-existing technology.

In more commercial sectors, external acquisition may include the acquisition of the technology firm that has the required products. Leading-edge technology organisations may well have an internal R&D function. This could take the form of a separate business unit within the main organisation. In a devolved management and accounting structure this unit might expect to have a quasi-contractual relationship with other parts of the organisation. Given the increasing cost of R&D work, strategic alliances are increasingly common. The key issue in any kind of R&D work is to ensure that all risks have been identified and the strengths and weaknesses in any proposed development measured and a response made.

Technology management

How do we define it?

There are several definitions of technology management. The National Research Council of the United States defined technology management as follows:

> Management of technology is a field that links engineering, science and management disciplines to plan, develop and implement technological capabilities to shape and accomplish the strategic and operational objectives of an organisation. (National Research Council, 1987)

More recently, the European Institute of Technology and Innovation Management put the following on its website:[1]

> Technology management addresses the effective identification, selection, acquisition, development, exploitation and protection of technologies (product, process and infrastructural) needed to maintain a market position and business performance in accordance with the company's objectives.

The management of technology is concerned with operating, improving and integrating an organisation's existing output, transformation and administration and control systems. It is also vitally concerned with the development and introduction of innovatory systems. Twiss and Goodridge (1989) list four areas of activity where technology can be applied:

- diversification;
- development of products or services for the existing business;
- production or operational systems;
- management systems.

Technology management is not something separate from other activities within an organisation, but should be the concern of everyone within an organisation, as technology is being applied to new functional areas, and the requirement for integration means that no person or department can remain unaffected or detached. As such, it is important to ensure that staff have the correct technology skills in order to make it work for the good of the organisation.

What are the prerequisites for effective technology management?

Technology management has a strong emphasis on the management of change, both technical and non-technical. Earlier in this chapter preparedness for devising and implementing an effective strategy was discussed. There is also a need to ensure that an organisation is ready for technology implementation. Twiss and Goodridge (1989) identify a number of prerequisites, within an overall framework that supports systematic change and encourages innovative development. These are as follows:

- sensitivity to trends in the total business environment;
- a long-term orientation;
- top management commitment to change;
- cross-functional integration;
- a high level of communications, both top-down and bottom-up;
- flexibility to enable a rapid response;
- an external orientation;
- creativity and a responsiveness to new ideas;
- the presence and encouragement of internal entrepreneurs;
- responsibility for all aspects of a change programme vested in one person;
- identification, capture and transfer of new knowledge;
- a focus on user needs and receptivity to user ideas;
- investment in education and training to support the change.

Twiss and Goodridge (1989) comment that 'the creation of the right environment for the generation and implementation of technical change to incorporate these features can be seen to be dependent upon a number of interrelating elements', listed as:

- corporate culture;

- the processes of strategy formulation and dissemination;
- the organisational structure;
- managerial information and control systems;
- the attitudes, motivations and contributions of individuals.

'If a change programme does not address all these elements it is unlikely to realise its full potential' (Twiss and Goodridge, 1989). The authors recommend an audit process to determine the readiness of an organisation for technology change. This should include (in the following order):

- analysis of the level of receptivity to change;
- assessment of the environment;
- identification of the extent of change required to meet objectives.

This process is normally carried out using audit techniques to assess culture, preparedness for innovation and present state of technology implementation.

Why do we need to manage technology strategically?

Early work by Schumpeter (1942) led to the widely accepted theory that the long-term performance of an economy depends on its ability to exploit successfully new technical possibilities. More recently, Stoneman and Vickers (1988) concluded that '... there is a considerable amount of evidence from different economies and different time periods to support the view that technological change is an important factor in the generation of economic growth. The estimates suggest that around 40 per cent of output growth can be attributed to new technology'. It has frequently been argued in recent years that the fundamental performance indicator of success for any organisation, sector or even country is long-term competitiveness. Hayes and Abernathy, quoted in Rhodes and Wield (1994), comment that 'success in most industries today requires an organisational commitment to compete in the marketplace on technological grounds – that is, to compete over the long run by offering superior products.'

Strategic technology management is a relatively new discipline, at least judging by the literature on the subject. As Coombs and Richards (1993) comment:

> The traditional approach to R&D management is being significantly expanded (in some firms) by a more ambitious function, which we can genuinely term 'strategic technology management'. In general, this function is higher up the corporate pecking order. Its essence is the management of a portfolio of technologies, in which the area of responsibility is the whole chain from acquisition, through development and the adding of a firm-specific competitive edge, to development for precise customer applications.

Where does know-how fit in?

It is not just the tangible benefits of the effective application of technology that make its strategic management an important means of adding value to an organisation. As technology becomes more pervasive, it stimulates the development of technology know-how – the knowledge possessed by the staff as to how to manage, develop and exploit the technology to the full. And it is arguably this intangible benefit that is most important to the future success of the organisation. As the European Union has recognised, it is the management of knowledge within technology situations that is of most importance:

> ... perhaps the biggest challenge presented by technology advance is not the acquisition of the knowledge itself, but

> rather a need for training in the management of technological change. There are many who fear that the limited understanding of many European managers of the nature of technological innovation and its human resources implications could become a main obstacle for successful adaptation of European industry to the changing economic and industrial environment. (Industrial Research and Development Advisory Committee of the Commission of the European Communities, 1991)

What does holistic, integrative management mean?

Many organisations that have invested heavily in new technology have been disappointed in the results. Why is this so? Twiss and Goodridge (1989) suggest that there are two main reasons:

- the strategic implications have not been fully understood;
- the impact of technical change on the organisation as a whole has been ignored.

Adopting new, better technology will not in itself solve an organisation's problems, give it a competitive edge or in some other way enable it to respond to both strategic and operational imperatives. Indeed, there are many examples where technology has actually disadvantaged an organisation. Technology managers need to have a broader perspective and a better appreciation and understanding of what success is and how it can be achieved, both with and without technology initiatives, before they embark upon new technology ventures. Too often the technology has led the organisation or project rather than the other way round. Strategy rather than technology-led initiatives are essential in this respect. These initiatives are ones that need to take due account of the technological options, but which also look at the challenges and problems in a holistic way.

For this reason, technology strategies must be linked in with the overarching corporate or business strategy to be truly successful. Without a serious, continuous, institution-wide corporate strategic planning process it will not be possible to determine any technology roadmaps effectively. It is the integration of technology and technology systems with the rest of an organisation's activities, policies, plans and strategies that is important. As Andrews and Stalick (1994) also comment, 'the sensible application of technology depends on the competent integration of technology with work processes'. Integrative technology management will form one of the key underlying themes of this book.

What are change and choice in technology management?

The global nature of technological change means that it is virtually impossible to control or channel technologies; it is more a question of managing their impact and integrating them to best effect within our organisations while trying to influence the wider future development of those technologies. Technology is becoming of increasing importance in the search for competitive advantage, improved quality, greater efficiency, increased effectiveness or indeed of all four combined. It is seen as something that can add significant value to a wide range of operations and activities. It makes the difference between competitiveness and uncompetitiveness; effectiveness or ineffectiveness; success or failure. 'Technology can be exploited to achieve corporate objectives in the context of the organisation's culture, managerial style, attitudes and management systems' (Twiss and Goodridge, 1987); 'technology affects the competitive position of the firm either through affecting its costs, or through differentiation'

(Porter, 1985) between its products or services and those of its competitors. Indeed, the (in)effectiveness of an organisation's approach to technology developments has the power to 'make or break' future success and a policy of resting on one's reputation without further innovation and improvement is not an option. 'The undisputed market and technology leader coasts on the momentum of past successes, failing to react strongly to the ... signals of impending decline' (Pascale, 1990). 'Getting to the future first ... [can provide] substantial rewards' (Hamel and Prahalad, 1994) although the risks are significant and need to be fully and properly appraised.

Strategic technology management, then, means taking decisions and making changes. A strategic approach is particularly recommended when new products, processes or services are being considered (Cooper and Kleinschmidt, 1987). One of the most difficult strategic decisions to take relates to technology choice. This is not just at the level of specific systems procurement, but relates just as importantly to the broader technology trends that affect the overall environment in which management decisions are made. Porter (1985) offers four scenarios as the basis for determining whether to embrace a technological change:

1. The technological change itself lowers costs or positively enhances differentiation, and the lead obtained is reasonably protected against rapid imitation.
2. The change favourably alters the costs or uniqueness of some part of the value chain. Imitation by other firms will still leave the innovating firm in a more favourable position than that before the innovation.
3. Pioneering a change may give the pioneer long-term advantages even in the face of imitators.
4. Technological change may favourably alter the structure of the industry.

These scenarios are all concerned with *differentiation*. Differentiation can of course relate not just to the product or service but also to the market or markets for which it is being provided. A number of possibilities are evident. A single, undifferentiated product could be provided to a single market, as, for example, with library services to an academic community. In reality, such a community or market is not homogeneous but is differentiated by different demands – academic staff, students, support workers, all of whom will be looking for something different from the service. Although they have different needs, they may be happy with the same, undifferentiated service (e.g. Internet access from the library through wireless technology) or they may require differentiated services, as for example from a virtual learning environment, where the lecturer will have different needs from those of the learner, even though the product/service is ostensibly the same. Sometimes, the market will have to be broken into *segments*. Library technology suppliers do not necessary supply to other markets, especially outside the public or educational sectors; libraries break down their services and their markets – or at least have done so traditionally. For example, traditional inter-library lending (ILL) was the preserve of academic staff in a higher education institution; thanks to technology developments, it is now an attractive service to all parts of an academic community, even though it is a demand coming from different segments within that community.

Whatever the market or the environment, in making a technology choice, strategy managers will have to ensure that there are sufficient advantages to the change and the technology being embraced to 'make a difference' and to justify the initial outlay. The investment may be considerable when it comes to significant invention or innovation, as discussed in Chapter 2. Even where it is more a case of improvement of existing systems – as is more likely to be the situation in LIS units – the investment is likely to be substantial in relation to the organisation's overall capital and recurrent budgets. Before even considering a major change, an organisation will need to be certain that it has sufficient resources available in order to ensure that the change proposed has the greatest chance of success once a go-ahead has been given.

What is a technology roadmap?

The options and choices facing the strategic technology manager are likely to require some kind of organisation or categorisation if they are not to bewilder decision makers. A technology roadmap can assist in both clarifying the position and highlighting and prioritising the realistic options as part of a strategy formulation process. Indeed, the completed roadmap document might usefully form a part of the final strategy document. The roadmap will typically comprise an organisation's plans for technology development as part of its future aims and objectives in terms of product or service delivery. It encourages the strategic technology manager to:

- anticipate future technological advances, including the prediction of appropriate technology life cycles;
- develop benchmark data;
- map out a pathway or trajectory for the organisation and its adoption or development of the chosen technology/ies;
- identify a programme or programmes of projects that will be necessary in order to adopt or develop the technology effectively and efficiently;
- include appropriate reference to financial, business and marketing requirements.

The ease or difficulty with which a reliable roadmap can be formulated will depend on the nature of the technology and its life cycle. If the technology is in its infancy, or has a long life cycle, then it will be difficult to determine how that technology will develop with any degree of accuracy except perhaps in the shorter term. The more developed the technology and/or the shorter the life cycle, the more the roadmap can be drawn up with confidence. The kind of forecasting that is needed to produce a roadmap is discussed in Chapters 3 and 4.

What does strategic positioning mean?

The end product of strategic technology management should be the identification, take-up and maintenance of an institutional position in relation to technology invention, innovation and improvement. This is the fundamental choice that every organisation – and not just the very active R&D-led units – must make. Although there is a broad spectrum of possibilities, there are basically three choices: pioneer/leader; fast follower; laggard. The position that an organisation takes may vary depending upon the particular environment, technology or overarching objective. Having said this, it must be recognised that there are some interdependencies that will make being a pioneer in one area incompatible with being a laggard in another: a library that wishes to be at the leading edge of digital resource provision, for example, is unlikely to be one that has yet to embrace Internet-based technologies more generally.

There are advantages and disadvantages to each of the three positions noted above. Being a laggard obviates the need for much R&D expenditure and the risk of failure is low because the technology being adopted will already be 'tried and tested'. However, the likelihood of significant financial, quality or service gain will be minimal because there will be little new about what is being offered once the technology choice has been implemented. Conversely, the investment and risk-taking required by a pioneer will be very high, but the rewards will be equally significant if the decision is a good one. If several pioneers are in the same environment and market at the same time, then there may well be an issue of standards – with a battle often ensuing until one dominant design or supplier wins. The classic example of this situation is the Betamax/VHS struggle in video-cassette recorders. A current LIS example relates to the question of dominant design in virtual and managed learning environment systems. A fast follower ends up somewhere in between these two extremes. Indeed, fast following may be the best option: the initial teething troubles have been solved, but the technology is still novel enough for its adoption to represent a major innovation and a point of significant differentiation from other competitors in the market. An alternative to these options is to contract out the function requiring a technology solution or the technology development itself. This may reduce the risk to nearly zero, but is also likely to reduce the significance of any gains that the organisation might hope to enjoy.

Summary

This chapter has looked at the fundamental aspects of strategic technology management. It has defined and analysed the three key aspects of strategy, technology and management and considered the main aspects of these terms as a preparation for the later, more detailed chapters.

Strategy is the output from a formalised planning process. It is characterised and codified through the creation of a strategic plan document. The key elements of such a document were described and discussed. Although the contents of a strategy may vary, it must state the strategic vision, the aims and objectives that will achieve that vision and the ways in which they will be met. The document should also refer to the organisation's current position and the views and requirements of the various stakeholders. It may also include higher level implementation milestones, risks and constraints. An effective and meaningful strategy is designed for implementation and capable of being implemented. It is flexible enough to be changed in the light of altered circumstances during the course of the planning period. Organisations will need to determine their strategic capabilities and the extent to which a strategy is realistic in their terms. A number of prerequisites were identified.

The transformation process in technology was described and a basic model of a technology system put forward. The process sums up the basic activity associated with technology applications. Technology trends were discussed and the effect of political and regulatory factors considered. The technology life cycle is an important consideration in strategic technology management, and technology substitution as a means of extending life cycles was discussed. Descriptions of the different kinds of technology, with special reference to critical, key, strategic and pacing technologies, were included.

The need to manage technology strategically was emphasised and justified. Increasingly, the emphasis is on developing long-term know-how that will transcend individual technologies in the context of a broad integration of technology management with an organisation's other strategies, plans, policies and activities. Strategic technology management requires change and choice to be embraced and decisions taken. The emphasis is likely to be on differentiation – of product, service or markets. A technology roadmap can lead the way to a strategic position with which the organisation is comfortable. The position can be one of three: pioneer/leader; fast follower; laggard.

Note

1. *http://www-eitm.eng.cam.ac.uk/purpose_mission.htm*

Bibliography

Andrews, D.C. and Stalick, S.K. (1994) *Business Reengineering: The Survival Guide*. New York: Prentice Hall.

Bart, C.K. (1998) 'A comparison of mission statements and their rationales in innovative and non-innovative firms', *International Journal of Technology Management*, 16(1–3).

Coombs, R. and Richards, R. (1993) 'Strategic control of technology in diversified companies with decentralised R&D', *Technology Analysis and Strategic Management* 5(4), 385–96.

Cooper, R.G. and Kleinschmidt, E.J. (1987) 'New products: what separates winners from losers?', *Journal of Production Innovation Management*, 4, 169–84.

Corrall, S. (1994) *Strategic Planning for Library and Information Services*. London: ASLIB.

Crawford, C.M. (1991) *New Products Management*. Homewood, IL: Irwin.

Feather, J. (2003) 'Theoretical perspectives on the information society', in S. Hornby and Z. Clarke (eds), *Challenge and Change in the Information Society*. London: Facet, 3–17.

Hamel, G. and Prahalad, C.K. (1994) *Competing for the Future*. Boston: Harvard Business School.

Hayes, R.H. and Abernathy, W.J. (1994) 'Managing our way to economic decline', in E. Rhodes and D. Wield (eds), *Implementing New Technologies: Innovation and the Management of Technology*, 2nd edn. Oxford: Blackwell.

HEFC (Higher Education Funding Council) (2003) *HEFCE Strategic Plan, 2003–08*. Bristol: HEFCE.

Klemm, M. et al. (1991) 'Mission statements: selling corporate values to employees', *Service Industries Journal*, 20(1), 22–39.

Mintzberg, H. (1994) *The Rise and Fall of Strategic Planning*. Englewood Cliffs, NJ: Prentice Hall.

National Research Council (1987) *Management of Technology: The Hidden Competitive Advantage. Report of the Task Force on the Management of Technology*. Washington, DC: National Academy Press.

Open University (1994) *T841: The Strategic Management of Technology*. Milton Keynes: Open University.

Pascale, R. (1990) *Managing on the Edge*. London: Penguin.

Pavitt, K. (1988) *Strategic Management in the Innovating Firm*, ESRC Paper 61. Essex: SPRU.

The Pocket MBA (1992). London: Economist Books.

Porter, M.E. (1985) *Competitive Advantage: Creating and Maintaining Superior Performance*. London: Collier Macmillan.

Rhodes, E. and Wield, D. (eds) (1994) *Implementing New Technologies: Innovation and the Management of Technology*, 2nd edn. Oxford: Blackwell.

Robert, M. (1993) *Strategy Pure and Simple – How Winning CEOs Outthink Their Competition*. New York: McGraw-Hill.

Schumpeter, J.A. (1942) *Capitalism, Socialism and Democracy*. New York: Harper.

Siess, J.A. (2002) *Time Management, Planning and Prioritisation for Librarians*. Lanham, MD: Scarecrow.

Snyder, H. and Davenport, E. (1997) *Costing and Pricing in the Digital Age: A Practical Guide for Information Services*. London: Library Association Publishing.

Steele, L.W. (1989) *Managing Technology: The Strategic View*. New York: McGraw-Hill.

Stoneman, P. and Vickers, J. (1988) 'The economics of technology policy', *Oxford Review of Economic Policy*, 4(4).

Stueart, R.D. and Moran, B.B. (1998) *Library and Information Centre Management*. Englewood, CO: Libraries Unlimited.

Twiss, B. and Goodridge, M. (1989) *Managing Technology for Competitive Advantage*. London: Pitman.

2

Invention, innovation, improvement, integration

Introduction

We now live in a world where innovation through technology invention, development and application is widely regarded as the norm. This situation can be characterised as not so much a series of step changes but a curve of continuous advancement and improvement. The growing importance of technology has been particularly noticeable in service industries and LIS is no exception, as discussed in Case study 1. This is because information creation, organisation, management, access and exploitation are a fundamental part of LIS work and technology applications have significant potential for improving all aspects of associated transformation processes.

This chapter is concerned with four basic concepts:

- *Improvement* – the enhancement, typically on a continuing basis, of existing systems, products and services to improve performance and hence effectiveness and/or efficiency.
- *Invention* – the development of a new idea or concept relating to a product, system or service; this may include some form of *prototyping*, where the new idea or concept is demonstrated in a practical form, though not necessarily in a way that can be marketed and sold.

- *Innovation* – the development of a product, process or service, typically from an invention, in such a way that it can be produced or provided on a commercial scale and exploited in mass markets.
- *Integration* – the bringing together of different technologies, processes, systems, activities, organisations or all of these to form a new product, process, service or technology, the aim being to progress by improving, innovating or even inventing.

There are close links between invention, innovation, re-innovation, improvement and continuous improvement. Evolutionary, incremental innovations and re-innovations cover the vast majority of technical change, especially in service industries such as LIS (Sundbo, 1997). Revolutionary or radical innovations account for the remaining 10 per cent (De Meyer, 1998). Improvement is about increasing efficiency or effectiveness. Innovation concerns the introduction of new products or services. This concept includes re-innovation and incremental innovation, where an existing product or service is redesigned to better effect. There is no clear division between the various categories of innovation and improvement; rather they are different stages on a continuum associated with change, development and improvement. At times, evolution may give way to revolution, as is the case with the reverse product cycle discussed later in this chapter. Whatever the change, it can affect input, transformation or output, or all three.

Because these close links exist, the importance of integrative management of technology has been recognised for some time. The methodologies described later in this book offer senior managers an opportunity to involve all types of staff in the development and implementation of technology, even though on a day-to-day basis there will be a greater or lesser emphasis on the operational or the innovative depending on the particular roles of those involved in any working groups using these methodologies.

The state of the art

Many writers have stressed the need for a full appreciation of the 'state of the art' when dealing with strategic technology management and technological innovation and improvement. It is interesting to note, of course that the word 'art' is used to refer to new technology. This state is also difficult to define at any one moment for the simple reason that, as already noted, invention and innovation in technological environments is moving so quickly that any description would be out of date by the time it was written.

There is also a need to differentiate between the *current* state of the art (that is, what the best organisations in the field are using or doing) and the *advanced* state of the art (what research and development and 'cutting'- or 'leading'-edge institutions are using or doing, and which may well become the current state of the art when the technology is adopted more widely). Case study 1 describes a situation where some LIS organisations 'led the way' in terms of adopting novel methods of document delivery. What was advanced state of the art in 1993 is now current state of the art. One of the key aspects of strategic technology management, of course, is to anticipate what future current states of the art will be and anticipate that altered state accordingly.

Why invent, innovate or improve?

The drive to invent, innovate or improve is driven by a whole range of reasons. Above all, however, the ability and capacity to innovate is fundamental to maintaining and developing competitive advantage. R&D is an integral part of many national, regional, sectoral and institutional agendas. The results of such programmes may stimulate the development of products that can be applied to the sector in which the original research programme was located, or more broadly. These products may themselves have been developed as a result of existing products' obsolescence. Fashions may change, perhaps driven by users' changing needs and wants. The need to improve efficiency and effectiveness is also an important driver. Examples of the imperatives that drive technological innovation are discussed in Case study 1.

The fast pace of technological change poses challenges for those heavily involved in invention and innovation in particular. As life cycles shorten, the ability to obtain a sufficiently rewarding return on investment is curtailed. At the same time, R&D costs are escalating. Developing new products means taking risks – not everything will succeed – and in an era when short-term gain or profit is important, invention and innovation may be difficult to justify. However, without it, organisations will stagnate and lose competitive position. Increasingly, R&D, at least at a pre-competitive stage, is done through partnership and collaboration. This has also been the case in the LIS sector, where programmes such as the Electronic Libraries Programme (e-Lib) allowed R&D within UKHE to take place in a way that would have been impossible if individual libraries had had to do the work on their own.

Landmark change, dominant, robust and lean designs

The scale of change as a result of invention, innovation and improvement can vary considerably. As already noted, the vast majority of change is incremental or evolutionary. From time to time, however, there are major, revolutionary, changes. They extend the current state of the art, as opposed to those incremental advances that merely modify it. 'Landmark' changes are few and far between, but when they do occur, they revolutionise the way in which we work or live. These changes create a 'paradigm shift' that may well create winners and losers in its wake. The development of Internet-based resources and services is one such landmark change or paradigm shift (Kingston, 2000), and provides a good example of major innovation within the LIS environment. The core technology is well known (electronic library developments are largely based on Internet and International Standards Organisation (ISO) protocols), although further development will require risk taking and market strategies. On the one hand, new technology is being applied to existing markets; on the other, new markets may be tapped once the products and services are fully developed.

A dominant design is just that: it dominates the technology, the markets and the competitive and collaborative environments; it provides the 'industry standard' until a landmark change delivers a new dominant design. There has long been a 'dominant design' in LIS – that of hard copy storage, access and delivery. Although there have been many process innovations over the years – notably in automation of library procedures – the sector is at the final 'specific pattern stage of innovation ...[and] vulnerable to the possibility of a revolutionary new product introduction' (Noori, 1990) – in this case the digital library. Within this new environment, existing suppliers and users may find it 'difficult to adapt to environmental changes with ... an ageing product' (Noori, 1990).

Designs are said to be 'robust' when they have sufficient capacity and flexibility to be adapted to a range of different sets of needs over time. Traditional library provision has remained remarkably robust in this context, at least in the higher education sector, where many researchers still value the collections of scholarly monographs in print-on-paper form. On the other hand, a 'lean' design may meet current user needs but is unlikely to be capable of expansion to cope with changed demands. Robust designs underpin successful 'core' products that can be varied depending upon the requirements of different market segments, as discussed later in this chapter.

Lean organisations and just-in-time delivery

A lean organisation is not the same as a lean design. Partly as a result of the development of business process re-engineering (BPR) techniques, discussed later in this chapter, questions were asked about the effectiveness and efficiency of holding a large 'inventory' (stock), initially in manufacturing industries. This led to the development of just-in-time (JIT) supply – known in Japan, from where the concept originated, as *kanban*. Instead of the large warehouse, materials and products were only brought in (often from third-party suppliers) just before they were actually needed. More recently, the concept of JIT has been adopted in service industries. The move from access to holdings in LIS units and the provision of document delivery services is a good example of this, as discussed in detail in Case study 1. This kind of improvement has been driven partly by the need to cut costs (for example in journal subscriptions) and partly by the development and application of technologies that have made it easier to offer remote document supply as a viable alternative to on-site provision.

Trajectories

Technologies and technology designs will each have some form of trajectory, along which it is possible to plot development and improvement over a period of time. Having said this, the trajectory is not predetermined, and can be affected by a wide range of factors that both constrain and liberate the product, process or service development and substantially affect its markets. The dominant features of a specific technology trajectory will vary over the life cycle of that technology. Initially, the emphasis is likely to be on functionality, where a new technology offers new features, systems or options that make it stand out from the competition and perhaps even allow it to change the current state of the art and the dominant design regime. Once the design framework and the market sector/market niches are established, the emphasis then normally turns to reliability and then to cost/price, before the product or service begins to fade into obsolescence and another trajectory begins. Organisations may follow particular technology trajectories depending upon their overall aims and objectives, developing or adopting specific designs or products as the basis of their own strategies. These trajectories are likely to encourage incremental development, though from time to time there will be a discontinuity as the dominant design changes and new trajectories have to be developed in its wake.

Invention and innovation

An invention is based on a new idea that is turned into some kind of conceptual model that demonstrates the feasibility of that idea. The typical output is some form of prototype or demonstrator. Although the prototyping may be successful, a good deal of work and financial outlay is still likely to be necessary in order to turn this 'proof of concept' into a working product or service that people want to buy or use (Pfeiffer and Goffin, 2000).

Innovation is concerned with the development and implementation of new systems, products or services and is typically based on inventions. Innovation has a number of aspects, as shown in Figure 2.1. It will be seen that the prototyping process is an important part of the move from research and development

Figure 2.1 **Innovation framework**

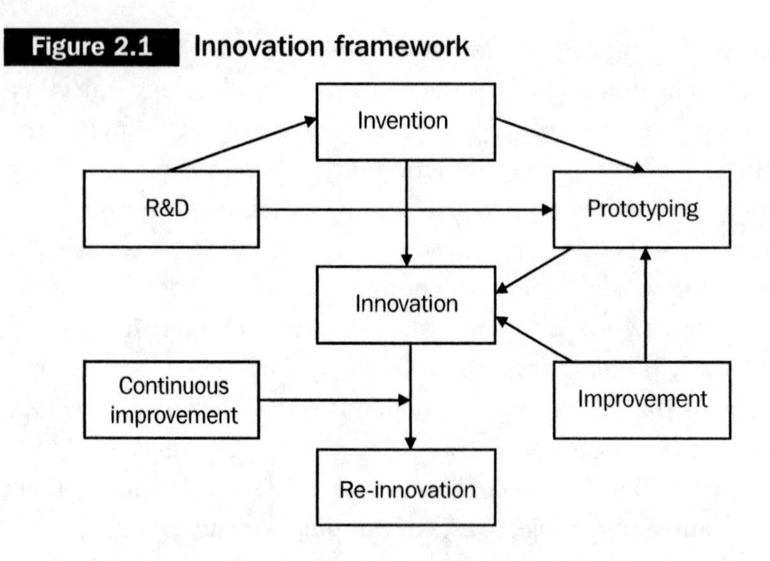

to marketable product or service. Innovation may have an invention at its root. In the case of LIS developments over the last five years, for example, the concept of the electronic library drove 'inventions', including protocols and standards for digital library application. The emergence of the digital library was also based on an innovatory change in perception: from holdings to access – the former previously a strength and the latter a weakness; more recently, just-in-time delivery, itself a 'crossover' technology concept from other industries and sectors, has been seen as the major strength, as discussed in Case study 1.

These inventions require research and development work to create them and then prototyping to provide practical demonstrations of the concept. At the prototyping stage there is typically still a good deal of work (and investment) to be undertaken. Radical innovation is not a neat process. It is often a messy business (Peters, 1988; Quinn, 1988) and controlling it is likely to mean that organisations are on the edge of their competence (Kantor, 1984).

Major innovation can be expensive (Kessler, 2000). As already noted, even large organisations are finding that they must collaborate in order to be able to afford the substantial capital costs associated with the development of new technologies and technology-based products and services. Significantly more expensive than the R&D endeavour is the commercialisation process by which a prototype becomes a product or a service. Nelson and Winter (1977) comment:

> Despite a tendency of some authors to try to slice neatly between invention, and adoption, with all of the uncertainty piled on the former, one cannot make sense of the micro studies of innovation unless one recognizes explicitly that many uncertainties cannot be resolved unless an innovation actually has been tried in practice.

But, as the authors point out, fashions can play an important part in the success or failure of an innovation and they are not always easy to forecast:

> Tastes matter; these may be hard to analyse and may not be stable. Even in situations where there is a relatively clear cut goal, and the decision to employ an innovation or not hinges on assessment of efficacy relative to that goal, it has proved hard to identify relevant criteria. (Nelson and Winter, 1977)

Feather (2003) adds:

> It is, by and large, users who determine the success or failure of particular systems and devices. Choices may be based on social convenience (the mobile phone), business efficiency (electronic mail), or economic necessity (financial information systems), but they are essentially choices. Technology does not determine what happens; it only determines what *can* happen.

The terms 'innovation' and 'improvement' need further interpretation in order to be fit for purpose in the service and public sectors. There is a degree of intangibility, simultaneity (i.e.

the new service is used at the same time as it is offered), divergence and obsolescence (e.g. journal half-life) about new and improved services that necessitates the creation of a definition of the innovation that allows the innovation to be effectively managed. This is especially true in the forward forecasting of reverse product cycles, discussed below and enumerated in Case study 1.

Re-innovation

Re-innovation can be defined as the redevelopment or design of an existing product, process, system or service to improve efficiency or effectiveness, often through re-engineering of the technology to produce cost savings. Innovation may come not through the development of a new technology but through the novel application of either existing or new technology-based products or services. Much of the technology that we currently use, and perhaps regard as innovative, is actually existing technology re-used or re-integrated in a novel way. This might take the form of re-innovation, where existing systems, processes or services are redesigned to reduce costs and improve efficiency and effectiveness, or both. Substitution innovation is about doing what you have always done, but doing it better through the application of technology, as was the case with the automation of libraries from the 1960s onwards. Feather (2003) comments:

> The society of the 21st century is critically dependent on information and communication technologies for a huge number of activities. Very few of these, however, are genuinely and inherently new. Some have displaced older systems, as paper-based correspondence has been largely, although by no means entirely, replaced by e-mail. Some operate in parallel to existing systems, as online databases sit happily alongside traditional reference books.

Application of innovation

We have seen that there are two types of innovation: radical and incremental. The former is likely to cost a good deal of money, take time and, if the application is successful, have a significant impact on the sector in which the invention is being applied and/or the innovation is being made. Incremental innovation, almost by definition, involves just the reverse. It carries much less risk and a significantly lower cost because of its step-by-step approach. Indeed, the effects of incremental innovation can often go unnoticed, at least in the first instance, although the reverse product cycle discussed below can often bring significant change, even though the early stages of this process may well not be detected and the later effects of the cycle cannot obviously be predicted.

Technology application, then, can often be a risky business. 'Even among the best performers, forays of more than moderate reach [in terms of innovation] quite frequently lead to problems' (Peters and Waterman, 1982). An innovation must be sufficiently attractive to stand a reasonable chance of successful adoption:

> A necessary condition for survival of an innovation is that, after a trial, it be perceived as worthwhile by the organisations that directly determine whether it is used or not. If the innovation is to persist and expand in use, the firm must find a new product or process profitable to produce or employ ... Consumers must be willing to buy the corn that the new seed produced at a price that covers cost. (Nelson and Winter, 1977)

In this context, additional functionality must be of paramount importance:

> Technology is essentially about function and performance. If one is not offering really new functionality or the potential for intrinsically superior performance, then success in the

> marketplace against established technologies is not likely ... the economic benefits of technological change are found in the marketplace. And the ability to plan the when, why and how of technological penetration into the marketplace is the key to making or losing money in the new technologies. (Betz, 1993)

Critical success factors

Later in this book the concept of critical success factors (CSFs) is discussed in relation to programme and project management. Reference is also made below to CSFs in the context of business process re-engineering (BPR). However, as noted above, CSFs are also important when it comes to the success or failure of a new product, service or other invention or innovation. Cooper and Kleinschmidt (1987) reinforce the conclusions of Betz and Nelson and Winter, quoted in the previous section, that 'product superiority' is the 'number one factor in success'. In other words, the new offering to the market gives customers or users something they did not have before that improved quality, reduced costs, solved problems as perceived by the customer/user or resulted in a product or service that was deemed to be superior to rival products or services.

Other key CSFs in the implementation of invention, innovation and improvement are likely to be: the proficiency with which the organisation implements its innovations, perhaps through the use of technology; a clear view of the product or service in relation to the target markets from an early stage in the development of the product or service; and a self-evident market potential for the product/service. The key CSFs will vary from industry to industry and sector to sector and later chapters of this book consider a range of techniques for identifying what they are likely to be. It should be stressed, of course, that the technology itself and/or its application is likely to be a/the major CSF to be considered. This is why technology assessment (TA), discussed in Chapter 3, is so important.

Business process re-engineering

A significant innovation typically requires a major, or even a radical, change in the way an organisation works, the goods that it produces or the services that it provides. Continuous improvement – discussed below – is insufficient to deliver such change. It may therefore require a re-engineering of the business processes that underpin the work of the organisation and, as a result, will need a significant and appropriate change in management processes to be in place if it is to be truly successful. The application of an innovation may represent a discontinuity within the organisation and this will inevitably need careful handling. The challenges that this brings are normally accommodated within a programme or project management structure, the success of which is crucial to the successful implementation of the innovation within a given environment. Success and failure in this area are discussed in Chapter 5.

BPR is a concept that was fashionable in the 1990s. Hammer and Stanton (1995) define it as 'the fundamental rethinking and radical redesign of business processes to bring about dramatic improvements in performance'. Hamel and Prahalad (1994) stress that 're-engineering aims to root out needless work and get every process in the [organisation] pointed in the direction of customer satisfaction, reduced cycle time and total quality'. In this context, it should be remembered that it is the process of transformation that is at the heart of technology and technological change. These transformation processes are at the heart of LIS work, whether it is issuing a book to a reader or providing a major digital library service. BPR is therefore a technique that deserves serious consideration in strategic technology management within the sector.

Although not written about as widely in recent years, BPR is still often adopted when there is a wish or a need to promote discontinuity as a part of a fundamental, innovative approach to the renewal of an organisation in order to accommodate major change – often brought about by technological development and implementation.

> BPR, by radically altering business processes and their accompanying organisational structures, fulfils ... [the] need [to] ... make substantial 'discontinuous' leaps in organisational performance ... Basically, BPR is taken to be an approach for generating radical improvements in the major dimensions that an organisation uses to compete in existing markets ... [it] provides radical improvements across a number of, if not all, competitive priorities. (Burgess, 1995)

In order to implement BPR, a number of stages must be completed. The process starts with the identification of the 'process vision' (Davenport, 1993; see also Phaal et al., 1998; Probert et al., 2000). This defines the process to be re-engineered, its objectives and its key attributes. Following on from this the main characteristics are identified: how will the process work? Leading on from this, performance measures and objectives are formulated: what does the process have to achieve? Perhaps the most fundamental stages are then reached: what are the CSFs and what are the potential barriers to implementation? CSFs and barriers to performance are discussed in more detail later.

Are BPR approaches valid in LIS organisations and public sector work more generally? In the private sector there is a clear objective: earning a profit. In the public sector, it is a case of a more broadly based concept of 'adding value'. But it is often difficult to be clear about what the value really is. Much depends upon the relative importance of the key stakeholders (discussed elsewhere in the book). For example, government may have a different view of the added value of LIS from the end-users. Discussions about BPR in a public sector context have therefore concentrated on 're-engineering for value' (Wreden, 1995), although, as noted by Halachmi and Bovaird (1997), it is often difficult to concentrate on either core business or core users when powerful stakeholders may insist on the continuation of marginal business at marginal value, not least because of political considerations.

The key point to stress is that BPR may be a necessary technique to use in the strategic management of technology, whether in the public or private sectors. As Andrews and Stalick (1994) comment: 'Historically, we too often have relied on technology to solve business problems. Overlaying technology on ineffective business processes only aggravates underlying problems. People then blame the technology and then the technology is not used effectively or even abandoned.' In other words, the technology may not be to blame. Nor is technology a universal panacea; it needs to be seen in context – as one way of innovating or improving. Other ways may well be found that do not centre upon technology but that improve efficiency or effectiveness in some other way. Combining a technology application and a BPR exercise may therefore yield particularly impressive results if properly integrated.

Improvement

In a competitive environment, organisations need to improve in order to maintain and preferably increase their competitiveness. In the public sector, this drive to increase competitiveness is typically translated into targets for increased efficiency and effectiveness. Technology is more and more seen as the means by which improvements can be made in order to achieve these aims. In a sense, these improvements represent incremental innovations.

There are various types of improvement that need to be considered. At the most fundamental level, changes may need to be made in order for a technological system to perform to its specification. In the manufacturing sector, this operational improvement is typically concerned with planning, monitoring and controlling activities that ensure any variations in system performance can be brought back or kept within agreed tolerances. These tolerances will certainly define a lowest acceptable performance level, but may also include a highest performance level, beyond which the system need not perform.

This concept can be translated across to service sectors such as LIS work. Take the example of a document delivery system where there is a lower performance specification of three days' turn-round time in terms of processing of requests and an upper performance specification of one day. A turn-round time of four days represents an occurrence out of the upper–lower limit of performance and requires corrective action. A turn-round time of half a day is similarly an out of limit occurrence and a decision has to be taken as to whether the 'improvement' in performance that this represents should be sustained through a revision of the performance specification. This tightening of the performance specification represents an operational improvement. It is characterised by an intention to improve (the intention typically being signalled by management), usually as part of a continuous improvement process within the organisation.

Continuous improvement

Continuous improvement (CI) is now a well-established concept, not least within LIS work. Approaches such as total quality management (TQM) and *kaizen* come under this heading.

Ultimately, the level of quality to be pursued and achieved is a strategic management decision. Once this is known, the organisation has to work out ways of ensuring conformance with that quality specification and CI should be part of a strategy that defines 'a coherent set of policies and a framework for decision-making in technology development' (Hill, 1985). These policies should be supported by benchmarking – a technique whereby other institutions within (or possibly also outside) the sector are used as comparators against which the organisation can test itself and identify areas where it needs to improve its own performance. Strategic benchmarking (in the form of the competitive or strategic web) is discussed later in the book. The important point to stress is that quality costs and that an understanding of this

> is essential for any business. The costs associated with the mismanagement of quality (the costs of non-conformance) are often large; are non-productive; and are avoidable through the implementation of TQM ... In most [organisations], the majority of quality related costs are incurred putting things right after they have gone wrong – the costs of non-conformance ... the objective of TQM is to continuously improve quality by eliminating non-conformance in every activity ... total quality involves everyone and influences the performance of the whole business. (Munro-Faure, 1993)

There are many techniques for the management of CI and the ways in which systems can be developed in order to ensure that products and services conform to the quality standards set. It is outside the scope of this book to look at them in any detail. Only one is discussed here as an exemplar (Crosby, 1979). Its main characteristics (which are similar to most quality management and improvement approaches) are as follows:

- there is an emphasis on 'conformance to requirements';
- there is a target of zero defects – perfection is possible;

- top management responsibility and attitude is crucial to success;
- an organisation's current quality management 'maturity' (cf. strategic IQ) and therefore its desired position regarding its approach to quality must be assessed;
- the cost of quality is used to measure the size of the quality problem;
- there is an emphasis on prevention rather than detection;
- there is a strong focus on changing corporate culture to ensure quality actually does improve.

This approach is set out in Figure 2.2.

Figure 2.2 Management of continuous improvement

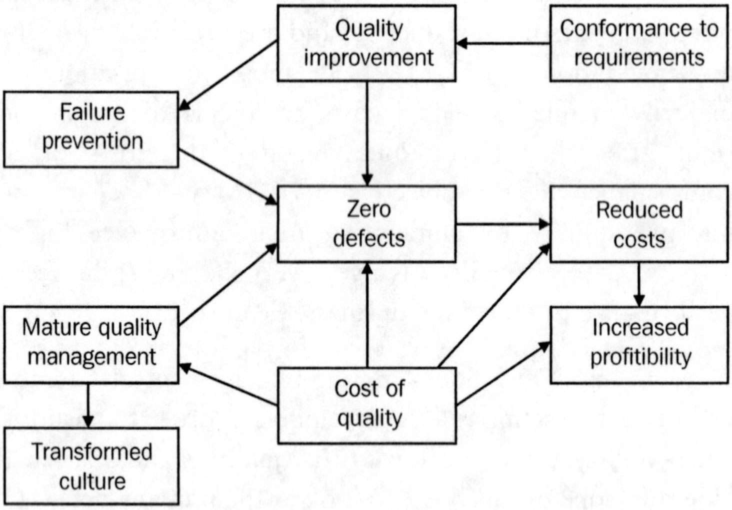

One important element of quality management, as noted in Figure 2.2, is the prevention of failure. This is an important aspect of strategic technology management, especially in terms of anticipating success or, perhaps more appropriately, possible blockages to it. One methodology that has a widespread validity in this context is failure mode and effects analysis (FMEA), discussed later in this book. FMEA is likely to be particularly useful for certain key aspects of strategic technology management because it identifies risk and the effect of proposed responses to risk. The methodology can point the way both to breakthrough projects (where significant change can bring about major quality improvements) and also to priority objectives for control in order to maintain existing quality levels.

Product development and reverse product cycle

In manufacturing industry, there has long been a clear model of a product development cycle, often known as the Abernathy–Utterback model (Abernathy and Utterback, 1978). This is summarised in Figure 2.3. The product development cycle is more difficult to apply in the service sector. Here, the concept of the reverse product cycle has been developed (Barras, 1986). This is summarised in Figure 2.4. Reverse product cycles have been evident in LIS work for some time. A particularly pervasive and influential example has been the development of innovative, and perhaps even revolutionary, means of providing library services through the development of digital delivery. This is explored in detail in Case study 1. The key point to stress with regard to the reverse product cycle is that the new product or service comes at the end of the cycle rather than at the beginning. The drive for improvement begins with increments and it is the cumulative effect of improvement and innovation that leads to the discontinuity of radical change through the appearance of a new dominant design.

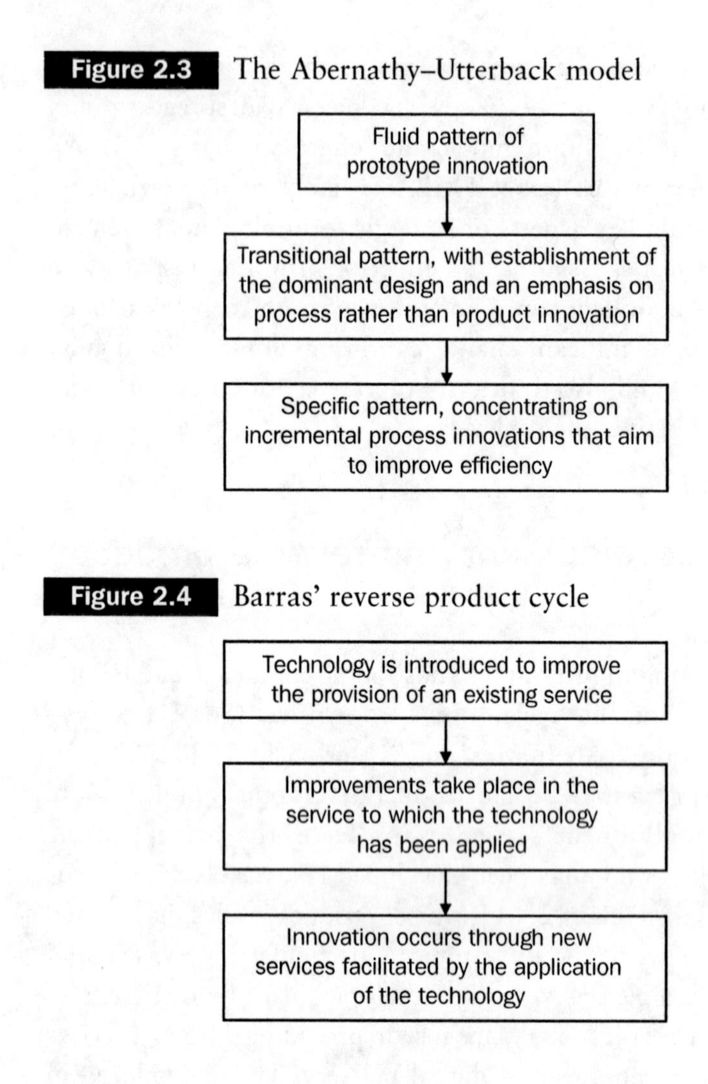

Figure 2.3 The Abernathy–Utterback model

Figure 2.4 Barras' reverse product cycle

Learning and experience curves

Given that one of the key reasons for improving performance is to reduce cost, there has been much interest in the learning or experience curve and the means by which that curve can be used to estimate the point at which maximum effectiveness and efficiency is reached and even to identify where and how it can be made to reach that maximum sooner than predicted. Learning curves are based on the premise that the more an operation is carried out, the more efficiently it will be done and so the more the cost of carrying it out will be reduced. Experience curves are concerned in addition with building capacity through the identification of economies of scale and making improvements across all parts of the organisation and its activities, whether in relation to an existing product or service using new approaches, or vice versa.

The steeper the experience curve and the shorter the time to market with either a new product/service or an existing product/service more efficiently produced and delivered, the greater the initial rewards for the organisation, though this has to be balanced against the possibility of competitors who then enter the market being able to leap-frog the original entrant. The great advantage for commercial organisations of forecasting and additionally influencing the learning and experience curves is that it allows them not only to anticipate the point at which maximum efficiency and effectiveness occur, but also to build their business and technology development plans on that forecast point and to price and estimate from the start on that basis.

Time (to market)

Time more generally is a crucial element in strategic technology management. The more innovative and novel a technology, the more time it will take to develop and 'productise'. Reference is made elsewhere to the significant demands placed on organisations engaged in significant R&D activity. Chapter 5, dealing with programme and project management, stresses the importance of time as both a resource and a challenge in terms of developing new products or services. Timing is also crucial: the point at which a new product or service is introduced can have a significant impact on its success. A launch that is too early may result in low take-up and withdrawal of the product or service because the market was not ready for it – perhaps the demand was not there or there were insufficient add-on benefits or support tools to make the product sufficiently attractive. On the other hand, launching a new product or service at a late stage in the development or life cycle of the underlying technology may be equally unsuccessful – too many other suppliers are already in the market, or the market itself is growing tired and looking for novel products or services.

Technology push and market pull

Two key terms in strategic technology management are technology push and market pull. Put simply, in the first, a technological answer is looking for a problem to solve; in the second, a market or markets are looking for a new product or service to solve a problem. In reality, the distinction between the pull and the push is not normally so clear cut as this. In Case study 1, for example, was it the availability of new technology or demands for new solutions to the problems of the information explosion and spiralling journal prices that fostered the development of the digital library? Both push and pull were involved here. In Case study 2, it is perhaps more a case of technology push not easily being matched by market pull. There are other drivers of technological development, application and change. Legislation and regulation can be important factors, especially with regard to market pull. Changes to systems may have to be made in order to comply with new laws (e.g. data protection or freedom of information).

Markets and market niche

The interaction between markets and technology is an important one and can affect the way an invention becomes a successful innovation. In this context, innovation is typically categorised into four different types:

- revolutionary innovation, where the technology and the markets are new;
- radical innovation, where the technology is new but the markets are the same;
- market niche, where the technology is not new but it is applied to new markets;
- regular innovation, which is, as already discussed, the further development of existing technology within existing markets.

Different sectors may foster one of these types of innovation over the others because of their characteristics and requirements. The LIS sector has a diverse set of requirements, but these are simple. However, the technology required to deliver them may be complex, as was the case with the development of electronic document delivery systems, noted in Case study 1. The basic search–locate–request–deliver model is an easy concept to understand, but it requires a substantial and sophisticated combination of technologies and standards to make it work well. Nor are users predictable: they have emotions as well as objectives and may have prejudices against or predilections for specific technologies. As Bitran and Pedrosa (1998) put it:

> The challenge is to design a service system for reproducibility but with flexibility to recognize and adapt to differences in culture, legal systems, regulations, and local infrastructure, among other factors.

On the other hand, simple and even insignificant changes to a product or service may make it possible for it to be used in markets other than those for which it was originally intended. This usually produces 'crossover' technology. This process typically occurs when the original product or service market is almost saturated and there is a need to diversify into other markets in order to maintain viability. Such changes can also, of course, be used to ensure that the product or service remains attractive within existing markets, at least until radically innovative ones can be launched. That is why we often see advertisements for 'new, improved' versions of existing products when, in fact, the changes made are mainly cosmetic.

Collaboration and competition

One important element of the broader environment in which strategy development takes place is the *industrial structure* and the *nature of competition* within that industry. Porter (1985) defines an industry as a group of firms that produce products that are close substitutes for each other. LIS is an industry in this context, because many library services can and do act as substitutes for each other. However, within the LIS 'industry', there are several different *sectors* (such as further or higher education) that serve specific markets and it is usually at the sector level where the environmental analysis is best performed. *Competitive advantage* is a topic more appropriate to the private rather than the public sector, though, as noted earlier in this book, there is an increasing degree of competitiveness even in areas where traditionally commercial values were not prevalent.

The nature of competition will vary from industry to industry and from sector to sector. In most LIS environments competition is likely to take the form of internal competition within the larger organisation of which most LIS units form a part or a sector – or subsector level of competition that relates to the benchmarking of services or the competitive bidding for funds. Determining competitive advantage requires an organisation to analyse and define its environment in terms of its competitors. This may be all other institutions in the same sector, but in reality it is likely to relate to a subset of organisations within that sector, or even ones outside the sector but which may have an interest in providing rival services, as for example with commercial (electronic) suppliers. Porter (1985) summarises the key determinants of 'industry competition' as those shown in Figure 2.5.

Figure 2.5 Porter's industry competition model

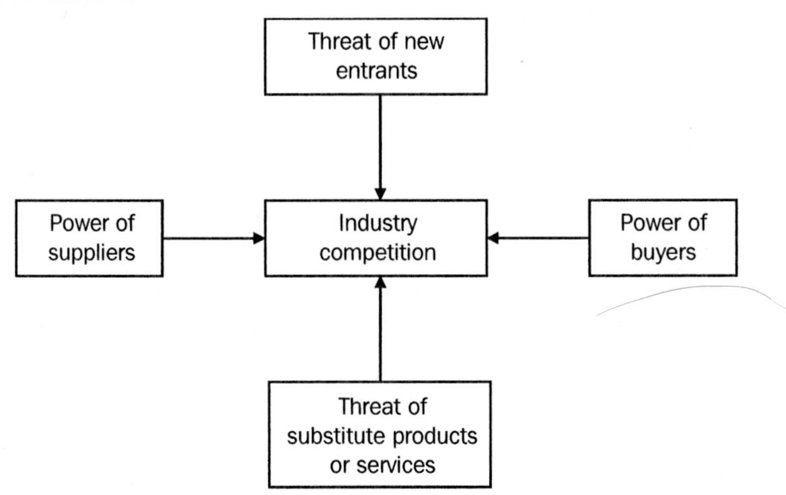

Activity within the LIS sector is likely to be characterised by a high degree of collaboration rather than competition – at least within public sector organisations – and a lack of 'differentiation' between the products or services offered. Indeed, the increasing trend towards benchmarking of LIS units is likely to lead to an increased uniformity of provision. Despite – or perhaps even because of – the high degree of consistency within the sector, there are threats of new entrants that need to be taken seriously by existing suppliers. New suppliers may not have to contend with the high fixed costs or the significant expense involved in, say, switching from hard copy to digital storage. Commercial suppliers of library materials may therefore be able to differentiate their product or service from those of the traditional library through advanced and innovative use of technology, though major innovations will require a high investment that may run the risk of not being returned sufficiently well to justify the initial outlay. This is where initial R&D work and subsequent prototyping of models through the many programmes of JISC in the UK has enabled a number of commercially viable 'spin-offs' to occur.

There are interesting parallels between private sector developments and the LIS sector. Over the last 15 years, there has been a move towards the 'extended firm', where communications networks in particular have enabled several organisations to work closely together to spread the risks involved in research and development through close strategic alliances. This approach has also become popular in the UK LIS sector, for example, with the strategic aim of cooperation between libraries, notably through the Internet, both to improve performance (as for example in respect of access to special collections) and to spread the costs of provision.

Make or buy

An organisation's technology strategy will provide a framework within which its choices regarding technology acquisition can be made. An important decision in strategic technology management is whether to develop a technology in-house, to buy it on the open market or to commission it from another organisation. In the early days of library automation, a number of libraries developed their own systems. A number of organisations in the UK, as for example universities, still develop their own software applications, especially in such areas as management information systems. However, recent practice has been heavily biased towards acquisition from commercial suppliers. In the case of new product development, the UK LIS sector has concentrated on partnership working with software companies, as was the case with the Joint Information Systems Committee's e-Lib programme and its successors. These various options can be summed up as 'make or buy'.

A key factor in determining whether to make or whether to buy is the *transaction cost*. Every transaction has a cost to it. Where a third-party supplier has been chosen, for example, there will be the cost of contract negotiation and transport/supply of the product or service that is the result of the contract. In-house software development will involve transactions between developer and client. The assessment of transaction costs will need to take into account the frequency and efficiency of current transactions, the extent to which they are routine or non-routine and certain or uncertain. The more frequent, routine and certain a transaction is, then the cheaper it is likely to be and vice versa. The presence or absence of an underpinning support infrastructure is also important. If this is present, then there is a built-in tendency to maintain the status quo of a transaction structure. If, for example, a LIS unit had contracted with a particular software supplier/developer and put in place the requisite infrastructure (including the necessary allocation of staff and their training for the transaction process), then the unit will be locked into a certain extent to that supplier rather than any other simply because the cost of transferring will be greater than staying with the present arrangement.

This is a frequent dilemma in procuring new library systems. Another example, at least in the past with regard to UK-based document delivery, would be a decision to switch from using the British Library as the main supplier. All the infrastructure has traditionally been set up to utilise the Library's services, and transaction costs can be minimised because there is both a well-established enabling infrastructure and a high degree of routine and certainty to the whole process. The more specific and uncertain transactions are, the more it will be that the whole transaction process takes place in-house rather than parts of it being outsourced to a third party. This is currently one of the factors that has to be taken into account when looking at the management of electronic document delivery services and their provision. However, as noted in Case studies 1 and 3, the high cost of R&D has meant that in public sector organisations at least, there is often little choice but to engage with third-party suppliers, even though there is a high degree of uncertainty and significant transaction costs are often involved.

Other factors need to be taken into account when considering a decision to make or buy. Making locks an organisation into a particular technology and timescale, whereas buying can offer the benefits of market developments. For highly competitive markets, of course, leading-edge proprietary development can help an organisation to become the benchmark developer within an area. Transaction cost analysis will tend to emphasise the short to medium term, whereas strategic technology management needs to take a longer-term view.

Integrative approaches

Over the last twenty years, integration has become a vital part of technology and technology management. The strategic alliances referred to above can become more closely integrative. Integration of technology development or usage is described as being either *horizontal* or *vertical*. When the integration is horizontal, it relates to a common approach or application across different parts of an organisation or service, as for example with an integrated library system. When it is vertical, it links together different kinds of organisation (supplier–library–publisher, for example) or different types of function and attribute (as for example with a technical service department and a bibliographical supplier). Integration is now widely practised on an international scale, not least by collective and shared use of the Internet, as for example with transatlantic virtual learning environments.

There is also the question of integration of innovation and improvement frameworks. As can be seen from the descriptions earlier in this chapter, there is a broad spectrum of change that ranges from the smallest development that comes about through a process of continuous improvement, through various forms of incremental innovation right through to radical change and a complete product cycle, reverse or otherwise.

Involvement in innovation and improvement

Innovation and improvement should not be something that happens separately from 'normal' work but should rather be an integral (as well as an integrative) part of everyday operations. This is certainly true of continuous improvement and re-innovation, but it is also applicable to more fundamental innovation. Innovation and invention are social processes (Burns and Stalker, 1961). People have to have a reason for inventing things or then applying those inventions and improving processes, products or services, and the reasons are typically driven by organisations, where the vast majority of inventions, innovations and improvements actually take place.

But there is another key stakeholder: the user. This is especially true in service industries such as LIS, where, as already noted, 'fashions' or client preferences can make a considerable difference to the success of an innovation or improvement. Self-service issuing, for example, is a technology that in many ways should be very attractive to users; however, the resilience of traditional issue and return, where a user transacts with a member of library staff (albeit with the record being created, altered or deleted using technology), has been significant and perhaps surprising. The user does not necessarily respond to the innovation in ways which those who propose it may think. The prediction of when a product or service disappears from the market is an especially difficult task, partly as a result of the 'vagaries' of user fashion. However, as explored in Case study 1, there will come a time when the producers of traditional products or services will find it nearly impossible to continue offering their wares as the dominant design changes and a majority of users or customers transfer to the new model, however strong their previous 'brand loyalty' actually was.

Innovative capacity and capability

Just as an organisation needs to be able to pass the 'strategic IQ test' described in Chapter 1, so must it be capable of innovating and improving in order to 'stay in business'. Strategic managers will need to determine not only the kind of innovation (as for example with radical versus incremental) that they believe is necessary for the organisation, but also the extent to which there is the capacity to embrace the chosen level of innovatory activity. The identification of capacity and capability is one dimension of strategy formulation and implementation, discussed later in this book.

Summary

This chapter has looked at the four themes of invention, innovation, improvement and integration. They are not separate activities, but form points on a continuum on which individual organisations will have to find their position (which may vary depending on the particular technology, product or service). An important concept in determining this position is the current and future state of the art and the ability to anticipate the latter rather than to respond to the former. The rationale for engaging with any of these activities is improvement in performance and competitiveness of the organisation.

Reference was made to the major or landmark changes that come about in technology, with existing dominant designs being replaced. In other words, the existing technology trajectories will be interrupted and new ones created. A robust design is likely to stand the test of time, whereas a lean one will only accommodate current requirements. Lean designs are not the same as lean organisations, where just-in-time concepts have come to the fore.

Invention and innovation go hand in hand. An innovation framework was described which linked R&D, prototyping, invention and innovation, in the context of (continuous) improvement. This may lead to re-innovation within existing products or services.

The application of innovation is challenging, and needs to take account of a number of critical success factors. One of these has to be the effectiveness and appropriateness of any business re-engineering process. Technology cannot itself solve problems, it can only assist in their solution; re-engineering may be the necessary fundamental shift that is required in order to ensure that the problem really is solved.

Improvement and continuous improvement are well-known concepts, underpinned by a long-standing TQM philosophy. In LIS units, the reality is likely to be one of reverse product cycles, where improvements lead to more fundamental shifts. This is explored particularly in Case study 1.

A number of other factors need to be considered. These were listed as: learning and experience curves where the steeper the curve, the quicker the time to market; time itself and its importance as a commodity as well as the criticality of timing; technology push and market pull. These factors all relate to markets and market positioning. Whether to compete or collaborate will be an important issue to resolve as will the decision to make or to buy technology as part of the strategic positioning drive. Involvement in innovation and improvement requires an integrative approach if it is to be successful, as also stressed in Chapter 1. It should also be a realistic approach, with the organisation's capacity and capability being taken into account.

It is important to stress that, although the various models described here seem to provide a series of clear trajectories where (a) invention is followed by innovation that is followed in turn by improvement or (b) there is a reverse product cycle beginning with incremental improvement and ending with major innovation or invention, the reality is much more complicated or diffuse. Only in retrospect does it seem to be clear what is happening. What is important is to try and forecast these trends and likelihoods before they happen in order to position the organisation to best effect. There is no single best way of managing invention, innovation or improvement. Different stakeholders will have differing perspectives, for one thing; the climate in which an innovation is to be made will also have a considerable effect on the success of the innovatory change. As noted earlier, this requires a high degree of risk management at all levels of innovation and improvement in order to maximise both the chances of success and the return on investment.

Bibliography

Abernathy, W.J. and Utterback, J.M. (1978) 'Patterns of industrial innovation', *Technology Review*, 80(7), 40–7.

Andrews D.C. and Stalick, S.K. (1994) *Business Reengineering: The Survival Guide*. New York: Prentice Hall.

Barras, R. (1986) 'Towards a theory of innovation in services', *Research Policy*, 15, 161–73.

Betz, F. (1993) *Strategic Technology Management*. New York: McGraw-Hill.

Bitran, G. and Pedrosa, L. (1998) 'A structured product development perspective for service operations', *European Management Journal*, 16(2) 169–89.

Burgess, T.F. (1995) 'Systems and re-engineering: relating the re-engineering paradigm to systems methodologies', *Systems Practice*, 8(6), 591–603.

Burns, T. and Stalker, G.M. (1961) *The Management of Innovation*. London: Tavistock.

Cooper, R.G. and Kleinschmidt, E.J. (1987) 'New products: what separates winners from losers?', *Journal of Production Innovation Management*, 4, 169–84.

Crosby, P.B. (1979) *Quality Is Free*. New York: McGraw-Hill.

Davenport, T.H. (1993) *Process Innovation: Re-Engineering Work Through Information*. Boston: Harvard Business School.

De Meyer, A. (1998) 'Manufacturing operations in Europe: where do we go next?', *European Management Journal*, 16(3), 262–71.

Feather, J. (2003) 'Theoretical perspectives on the information society', in S. Hornby and Z. Clarke (eds), *Challenge and Change in the Information Society*. London: Facet, 3–17.

Halachmi, A. and Bovaird, T. (1997) 'Process re-engineering in the public sector: learning some private sector lessons', *Technovation*, 17(5), 227–35.

Hamel, G. and Prahalad, C.K. (1994) *Competing for the Future*. Boston: Harvard Business School.

Hammer, M. and Stanton, S.A. (1995) *The Re-Engineering Revolution*. London: Harper Collins.

Hill, T. (1985) *Manufacturing Strategy*. London: Macmillan.

Kantor, R.M. (1984) *The Change Masters*. London: Routledge.

Kessler, E.H. (2000) 'Tightening the belt: methods for reducing development costs associated with new product development',

Journal of Engineering and Technology Management, 17, 59–92.

Kingston, W. (2000) 'Antibiotics, invention and innovation', *Research Policy*, 29, 679–710.

Munro-Faure, L. M. (1993) *Implementing Total Quality Management*. London: Pitman.

Nelson, R.R. and Winter, S.G. (1977) *An Evolutionary Theory of Economic Change*. Cambridge, MA: Harvard University Press.

Noori, H. (1990) *Managing the Dynamics of New Technology: Issues in Manufacturing Management*. Englewood Cliffs, NJ: Prentice Hall.

Peters, T. (1988) 'The mythology of innovation or a skunkworks tale', in M.L. Tushman and W.L. Moore (eds), *Readings in the Management of Innovation*. Boston: Pitman.

Peters, T. and Waterman, R. (1982) *In Search of Excellence*. New York: Harper & Row.

Pfeiffer, R. and Goffin, K. (2000) 'Getting the big idea', *The Engineer*, 4 February, 22–3.

Phaal, R. et. al. (1998) 'Technology management in manufacturing business: process and practical assessment', *Technovation*, 18, 541–53.

Porter, M.E. (1985) *Competitive Advantage: Creating and Maintaining Superior Performance*. London: Collier Macmillan.

Probert, D.R. et al. (2000) 'Development of a structured approach to assessing technology management practice', *Proceedings of the Institution of Mechanical Engineers*, 214(B), 313–21.

Quinn, J.B. (1988) 'Managing innovation: controlled chaos', in J. B. Quinn et al. (eds), *The Strategy Process*. Englewood Cliffs, NJ: Prentice-Hall.

Sundbo, J. (1997) 'Management of innovation in services', *Service Industries Journal*, 17(3), 432–55.

Wreden, N. (1995) 'Re-engineering for revenue', *Beyond Computing*, 4(7), 30–6.

3

Strategy formulation I: scenario planning

Introduction

The next two chapters are concerned with the techniques that can be used to formulate strategy. Many of these techniques are common to all forms of strategic planning and management. It should be stressed that many techniques have been excluded from this book; only those that seemed particularly relevant to technology management in a LIS context have been included. In any case, no single technique will solve all problems; there is no panacea to an area of management that becomes more difficult the more radical and innovative the technology being introduced. However, it is argued that a sensible and appropriate combination of these methodologies in an integrated way will provide the best response to the strategic management of technology.

There are several key benefits to the proper use of the techniques described in these chapters:

- The quality of decision making is improved.
- Opportunities and problems can be identified – and dealt with – effectively and efficiently.
- An appropriate view of the issues can be taken – a broad view in the case of strategy development, a focused one in relation to strategy implementation.

Forecasting

Within the last 10–15 years, the concept of *foresight* has become increasingly important as the basis for effective strategic technology management. This requires the integration of technology forecasting with an analysis of broader environmental issues so that strategic decision makers are 'informed by deep insights into trends in lifestyles, technology, demography and geopolitics' (Hamel and Prahalad, 1994). This has also been recognised by Feather (2003), who identifies four broad frameworks within which the 'Information Society' can be analysed: economic, technological, sociological and historical. Fashion may also play a part in determining future strategy. To what extent, for example, was the predominance of mobile phones among younger members of the population in the UK the result of a market pull driven by fashion as much as innovation?

Where the organisation is 'technologically active', it is of paramount importance that likely future developments are foreseen in order that an effective response can be made. Even where technology is not seen as being a key driver of an organisation's future strategy, it is important to ensure that any possible environmental impact is noted; any organisation can be taken unawares, as Case study 1 suggests. The more innovative and radical a new technology is, the greater the risk inherent in its adoption, especially in the early stages. Risk management, then, is a vital response to the results of a technology forecasting exercise, as discussed later in this book.

There are several key aspects that need to be investigated in forecasting future scenarios in respect of technology development in general. Twiss and Goodridge (1989) suggest that these are:

- appropriateness of the timescale for the analysis;
- significance of technology trends within the sector within the given timescale;

- effectiveness of the evaluation methodology/ies regarding the identified trends;
- likely consequences of the trends for the structure and capabilities of the industry/sector, the shape and nature of the market and the needs of the customers/users.

Technology forecasting is an art as much as a science. It is necessary to stress that just because one has made predictions – on which it is then proposed to build a strategy – does not mean to say that those predictions will come true. In looking at the techniques discussed below it is important to remember that the results of using them create one or more scenarios – and they are no more than that until the point is reached at which they become reality, or not, as the case may be.

The key challenge for the strategic technology manager will be to predict likely future political and regulatory priorities and to anticipate them as appropriate. So, for example, a LIS unit that is keen to bid for government funds to engage in technology development or application projects will need to determine the nature of forthcoming political correctness in its particular areas of interest. Caution needs to be exercised, of course, in that governments fall from power or stay in power but change their minds because of developments in the political climates in which they operate. Prediction is made all the more difficult because of this. In addition, by the time a government or other regulatory body has developed a framework within which a technology can either be supported or regulated, the technology may have 'moved on' to a point where the political initiatives if not the underlying imperatives are no longer relevant.

Environmental analysis

Organisations have to consider a large number of environmental factors when they are considering future strategy in general and their ability to compete in particular. One set of environmental factors relates to the broad 'climate' in which the organisation is working. The other set relates to the products or services that form the organisation's output.

Fahey and Narayan (1986) discuss the main aspects of *environmental analysis*. They define four stages: scanning, monitoring, forecasting and assessment. Scanning and monitoring the environment 'identifies surprises or strategic issues requiring action on the part of the organisation'. It is not just about an analysis of the current position but is more appropriately concerned with a study of present and future *trends*. There is therefore widespread agreement that the process has to be as broad as possible, as the early hints of major change may not always come from the obvious sources. Use of the Delphi technique can help here; it can also assist in the integration of different sources of data into a meaningful whole.

Delphi

Delphi is a qualitative method of forecasting that employs a team approach to decision-making. However, the team does not have to be assembled in one place, which means that the number of members can be much greater than you might normally consider possible. It is a method of developing expert consensus about a topic through a series of anonymous mailed questionnaires. The Delphi method has been employed in technological forecasting, planning and a variety of other areas.

The main stages are as follows:

1. Develop a scenario showing the problem/opportunity situation and the general reason for concern.
2. Select the team to be questioned. Team members must:
   - have a sense of involvement in the situation;
   - be in possession of relevant information;
   - be motivated to spend time on the Delphi process;
   - have a perception of the value of the information they will obtain from the other participants.

   (The team may vary in size. The size of the group can be dictated partly by the type of question being asked. For technological forecasting, 15–20 specialists in the area should be sufficient.)

The initial questionnaire should pose the scenario to the respondents with the general reason for concern. Then there should be two or three open-ended questions and/or requests for examples. There should also be a covering letter explaining the purpose of the questionnaire, the use of the results, instructions and a response deadline. Follow-up letters should be sent to non-respondents. Analyse the results of the first questionnaire. The aim of the analysis is to summarise all the responses in such a way that they can be clearly understood by the respondents in the second questionnaire. A good way of doing this would be to use an affinity diagram.

Develop a second questionnaire using the responses summarised from the first. The questions should focus on:

- identifying areas of agreement and disagreement;
- providing an opportunity to clarify meanings;
- establishing tentative priorities for the topics or solutions.

The aim of the questionnaire is to obtain clear responses that could be construed as a 'vote' for one type of decision. This questionnaire should first be tested on a non-team member before being sent to the respondents, again with a clear deadline. Analyse the second questionnaire. This involves mainly counting 'votes' and displaying them as a bar chart. There should also be some analysis of comments.

You can now decide whether to terminate the exercise or continue. This depends on whether the results are clear enough to formulate a decision or whether some of the comments raise new ideas that need to be explored further. If you continue, the third and last questionnaire must pull together the entire Delphi process. Do not omit questions because they have been asked before. The results of the second questionnaire should be summarised and sent to the participants. Ask the participants to vote on or rank the items, and remember to test the questionnaire on a non-team member before mailing it. Analyse the third questionnaire and draw conclusions as to the real problem or opportunity or possible routes to your objectives. Finally, report the results of the questionnaires to all the participants. The report should review the original situation, goals or process, procedure used and final results, and, if possible, any decisions made.

Below is a question from the first questionnaire for a Delphi study carried out at the University of East Anglia (UEA), together with the summary of the third and final Delphi survey question that resulted.

UEA NORWICH: DEVELOPING OUR FUTURE INFORMATION STRATEGY AND PROVISION

DELPHI QUESTIONNAIRE I

Please comment on the following statements and respond to the following questions in your own words. The aim of this first questionnaire is to identify the main themes from which we shall be developing more focused questionnaires and, eventually, a new Information Strategy for UEA. Please be as full in your answers as possible.

1. **The Context**

1.1. The ISD's [Information Services Direcorate] strategy group believes that academic support services can only operate within the context of their local institutional context. This means that whatever corporate approach is adopted by UEA will have to be followed by the ISD and reflected in the university's Information Strategy. However, UK higher education is changing, and it would be helpful to know what you see as the main aspects of UK HE development over the next 5–10 years which you believe will impinge not only on universities like UEA, but also particularly on academic support services, as represented at UEA by the ISD. In particular, comments will be welcome on the way in which teaching and learning, research and management activities will change and develop over the next 5–10 years.

Please list below, then, what you believe these aspects will be and on how teaching/learning, research and management activity within UK higher education will change:

1(a) THE CONTEXT

Respondents were asked to say what they thought would be the main contextual issues, which would impinge on the development of an Information Strategy over the next 5–10 years, with special reference to teaching and learning, research and management activities. No major discontinuities were identified, but a number of major trends emerged, as follows:

1.1. Income streams will continue to diversify within universities as public funding of HE reduces in real terms. There may be more changes still to come in the relationship between HE and FE funding (especially as vocational emphases increase), with government pressure for greater integration. There will be a closer coupling between HE and employment/employability. There will also be far more accountability (e.g. teaching and research assessments, value for money) and tighter governance required by government. There will be a further increase (unquantified) in the numbers of people entering higher education within the UK. Students will increasingly pay for their education and be more demanding 'customers' as a result. The possibility of differential fees after the next election might create greater student interest in services additional to the core of teaching and research.

1.2.1. There will be a greater integration with commerce and industry. There will be direct sponsorship through the commissioning of research and teaching and the enrolment of part-time, mature and sponsored students with ongoing workplace links. Commercial exploitation of the fruits of research will be a major reason for committing research investment. Private sector involvement in the selection and funding of research programmes will increase markedly (and not necessarily for the good – the involvement of more private money in research will need careful monitoring). A global HE economy will develop.

1.2.2. There will be an increasing regionalism in the organisation, management and take-up of university education. The regional agenda of government, e.g. through RDAs [Regional Development Agencies] will be important, although regional leadership is difficult to predict/prescribe as 'regions' vary so much in size and character. RDAs have a requirement to offer regional networks – can this be delivered in conjunction with HE, e.g. through MANs and RANs [metropolitan and regional area networks]?

1.3.1. It is most unlikely that 130 or so HE institutions will continue to survive independently. There is likely to be a considerable rationalisation between universities, facilitated by HEFCE, and/or possible federal structures or mergers bringing together universities into joint management arrangements.

1.3.2. There will be a slow, but continuing emergence of centres of excellence for research *and* teaching in particular subject areas, perhaps polarising the characteristics of individual universities. There will be a gap between an English version of the 'Ivy League' and other UK universities.

1.3.3. Inter/cross-disciplinary studies will grow in importance but medium-sized HEIs will have to focus their efforts on a manageable range of subjects to be cost-effective, aiming to find niche markets for teaching and learning and to achieve international excellence in research. Publication via electronic media/Internet texts will become increasingly more accepted as conferring research status. National leadership is required to determine a national strategy for supporting research, but it may not be given. While national strategies may be set in place, it will be following them that matters: what will be the carrots? What price institutional autonomy? Research activity will need to be protected.

1.4.1. Boundaries between taught postgraduates and undergraduate studies will become ever more blurred in many subjects, though the distinction between research-based postgraduates and taught ones will remain. There will be a greater emphasis on lifelong learning (including access to 'partial' qualifications – i.e. diploma not degree), with pressure from government to move towards flexible and distance approaches.

1.4.2. There will be a reduced reliance on the calendar to dictate the framework of the study year. There will be a tendency for telecommunications to increase the proportion of remote students at both undergraduate and graduate levels, with increasing access from abroad to remotely accessed UK courses, and vice versa. There will be increasing moves to share resources for teaching and learning across institutions for reasons of economy, including the possible provision of modular degrees from more than one HEI. This resource sharing may even extend beyond HEIs (e.g. to FE colleges). Open or flexible learning will require the association of resources with the curriculum in an atomic manner. This will entail a greater use of buildings via out-of-traditional-hours activity, with all its implications for security controls etc.

1.5. Users will demand increasing levels of integration between electronic services and hard copy availability. 'Library' material will need to be made easily available to remote students, although network managers will be forced to strike a balance between ease of (remote) access and adequate security. Twenty-four-hour service provision will be needed. Teaching material will be put up on the Intranet and the Internet will be widely used. The Library/ISD may become a gateway mediating the appropriate source of resources – internal or third party, physical or electronic. Hybrid services delivering 'new' as well as existing media will be required and the impact on traditional services and collections (notably journals) will have to be managed. Internet (and perhaps Web/digital TV) in the home will have an impact. HE will need to be careful not to exclude access to courses and materials incorrectly.

1.6. Allocation of budgets to ISD-type units will be linked more closely to perceived value of output. As value of output correlates closely with availability of electronic services, those institutions which invest most in electronic services and delivery will enter a virtuous circle. Within HEIs, departmental heads and even individual researchers will have an increasing role in content and service selection. Efficiency needs will drive centralisation of control but not of delivery, while central IT/IS departments should begin to have more control over budget allocation and content acquisition. There should be a central IT strategy (including purchase of hard/software) for basic use. Specialist needs should be funded and supported by schools as *agreed* exceptions.

1.7. The pace of change will continue to be rapid, requiring quick, flexible responses to opportunities (and threats). This will put more pressure on managers at all levels, making management competence and management information even more important. The role of librarian will change to that of 'gatekeeper' to information sources and purchasing/rights manager, and will be increasingly integrated into the IT department. The administration will become part of a service offering education to people as a consumer good. Standards of performance will become a critical part of attracting students.

Portfolio analysis

Most organisations supply or provide a range of products or services. These products or services will vary from the very profitable (as defined by the organisation's business or financial strategy) to the not-so-profitable or loss making. Products or services in the early stages of development are likely to fall into the latter category. These may be cross-subsidised from the more lucrative offerings until such time as they can stand alone financially. At the same time, an organisation cannot rely too much or for too long on its 'cash cows' – the income earners that support the activities that do not recoup their costs. They may need re-investment, renewal or re-engineering and, at some point, will come to the end of their natural life cycle. There is therefore a balance to be struck in providing a portfolio that allows for new products or services to be developed, but not wholly at the expense of the current money earners and their future potential.

A number of techniques are available to support the analysis of current and possible future portfolios. These are typically based on a matrix approach, where the products or services are plotted against two axes. A simple model uses rate of growth and market share relative to the market leader as the two variables (see Figure 3.1). Alternatively, the axes may be the attractiveness of the product/service or the market and the competitive position of the organisation.

Figure 3.1 Simple model for portfolio analysis

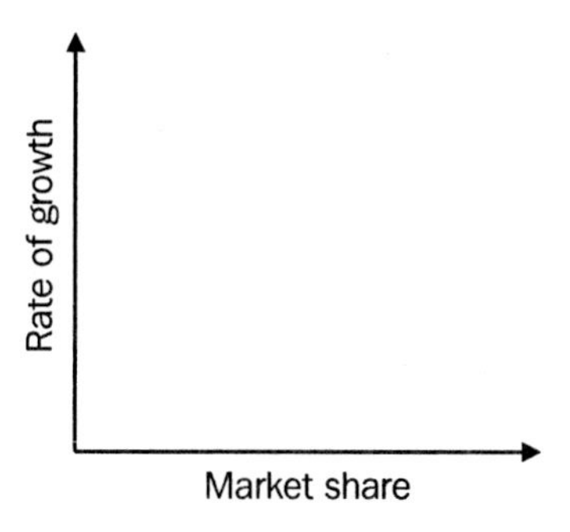

These portfolio matrixes are arguably more relevant to the private rather than the public sector. However, as the public sector now operates in a more competitive environment, these approaches may have their uses. Certainly it is the case that academic organisations, for example, undertake portfolio reviews that take into account the kinds of matrix headings described here. LIS units, too, may find the approach useful in terms of analysing the present and possible future mix of services that should best be offered, although the variables are likely to stress the demand for/popularity of particular services and their cost-effectiveness relative to identified benchmarks rather than strict market share or profitability. In identifying the strong and weak performers in a LIS unit, there will be an opportunity to determine how technology applications might either reinforce a good position or improve a weak one. In carrying out this analysis, of course, LIS unit managers will need to recognise that although a service may be unprofitable in strict financial terms, it may be academically or socially profitable, or a prerequisite of funding or constitution, regardless of any diseconomies of provision.

Market analysis

The above discussion about portfolios suggests that there is a strategic relationship between products or services and markets. Market analysis is the other side of portfolio analysis. In other words, whereas portfolio analysis looks particularly at what is provided in relation to the market, market analysis considers what the markets are and what they require, whether or not it is currently provided. The distinction is a fine one and, in reality, neither the market nor the portfolio would be looked at in isolation. However, it is perhaps helpful to consider the ways in which a strategic technology manager might usefully look at the markets in which the unit is operating.

In carrying out a market analysis, it is assumed that the key objective is to test the potential for growth within the identified priority markets and/or to assess the viability of remaining within a market, whether at the present or some other level (higher and lower) of activity. Kotler (1988) talks of three kinds of market growth: intensive, integrative and diversified. Intensive growth will relate largely to existing activity and could encompass the identification and development of new markets for current products and services (such as a public library intensifying its provision of services to HE students), increasing the share of existing markets or the development of new products or services for those existing markets (as for example with electronic document delivery).

Integrative growth, in Kotler's model, relates to merger, acquisition or some other form of control over businesses that are related to, but not currently part of, the organisation's activities. Three forms of integration are common. Backward integration relates to suppliers, forward integration to customers and horizontal integration to competitors. The company Ingenta[1] is a good example of a commercial organisation that has grown through this kind of integrative process.

Diversified growth, on the other hand, is about acquiring or developing businesses that are unrelated to existing activities. Kotler lists three kinds of diversification. Concentric diversification concerns developing new products or services that are attractive to new customers but which have strong links with existing products or services (as for example with distance learning); conglomerate diversification relates to the acquisition or development of activities that are unrelated to existing activities (such as a library acquiring a restaurant); and horizontal diversification concerns new developments that appeal to existing markets but that have no links with existing products and services or the technology that they use (as for example with a study skills advisory service).

Although market analysis may not seem a particularly valid or valuable approach for LIS units, it should be remembered that the vast majority of technology suppliers will be carrying out these analyses, and to be aware of their philosophy and strategy should prove useful when in negotiation with them with regard to the provision of new systems, products and services.

Product development mapping

Those industries that develop, market and sell products normally use formal product development mapping procedures in order to help them identify and exploit both new and existing markets. Given the fast-moving nature of technology development, this mapping process has to be continuous. It is also a useful tool for those who buy those products or who otherwise need to be aware of the interaction between product and market. Reference has already been made to the use of prototypes to move an invention or an innovation to a saleable product or service. Mapping begins in earnest at the point where there is a 'core' product that can be developed to satisfy different markets' needs. These developments typically result in a number of variants of the core product, as shown in Table 3.1.

Fusion is arguably the most important category in terms of strategic technology management:

> The fusion of technologies goes beyond mere combination. Fusion is more than complementarism, because it creates a new market and new growth opportunities for each participant in the innovation. It blends incremental improvements from several (often previously separate) fields to create a product. (Kodama, 1991)

Table 3.1 Core product variants

Product or service variant	Description in relation to the core product
Enhanced	In this scenario, additional, distinctive features are added to the core product to make it more attractive to a particular market or market segment. A generic system might be adapted for the particular requirements of academic libraries, for example.
Customised	The core product is developed and enhanced for a specific customer in order to meet their additional requirements. This kind of enhancement may include some of the features of the more 'mainstream' enhanced version or versions. In LIS work this might be the adaptation for a large or very specialist client where, although the core functionality might be acceptable, the requirements are on such a scale or of such distinctiveness that one-off or narrowly defined enhancements have to be made.
Commodity	The core product is stripped down to its basics in order to reduce the price and offer a version 'without the frills' that may appeal to a mass market or a market segment that cannot afford the full 'core' product, let alone the enhanced versions. An example in LIS work might be offering a basic housekeeping system for smaller libraries.
Hybrid	Here, two core products are typically combined to provide a new product. The development of the hybrid library has been one of the major changes in LIS work in recent years. It builds on the best of the traditional dominant design of library provision while exploiting all the advantages and opportunities presented by Internet-based and related technology.
Fusion	The fusion of technologies sees a fundamental joining of two or more technologies and develops new markets in the process. There will typically be new linkages between industries that did not previously exist even in a hybrid form.

It has been particularly prevalent in service industries, such as LIS, where information technology has been used to replace paper-based systems. A company like Ingenta, for example, has not only fused a number of technologies together, but has also created those novel linkages between different industries (and notably publishers and document suppliers) that are such a fundamental part of technology fusion.

Quality function deployment

Product mapping is intended to identify the key technology and market drivers of product or service development and helps managers to identify those areas where there is likely to be the greatest return on investment. In the context of strategic technology management, this approach is developed later in the book into the strategic or competitive web approach. In more commercially orientated organisations, there are several techniques that aim to bring together product or service development in order to map the areas where an organisation can best compete and/or where it should be most engaged in collaboration with others. The example given here relates to the technique called quality function deployment (QFD). QFD:

- covers the whole of the 'production' process;
- aims to translate customer requirements into organisational requirements;
- orders the importance of requirements;
- focuses on product and process rather than people;
- uses multidisciplinary teams;
- assumes 'hard' statistical and analytical tools are already in place;

- moves from 'wants' to 'hows' at each stage of the production process;
- identifies the relative strength of relationships;
- results in quantitative targets;
- requires time.

It begins with the identification of customer requirements and ends with the development of a product or service that meets the needs of the market for which it was intended (see Figure 3.2).

Figure 3.2 Quality function deployment

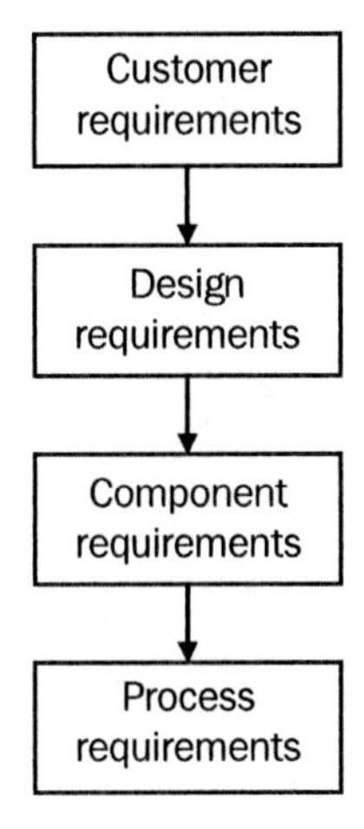

Stakeholder analysis

Reference has been made in earlier chapters to the relationship between strategy and stakeholders. Organisational strategies typically refer to those individuals, groups, agencies or institutions that have an interest or 'stake' in the organisation, its direction and its future success. Some stakeholders may be able to affect the strategy; others may be affected by it. An academic library is typically 'governed' by some form of user committee, where teaching and research staff are likely to be in a majority. They can often affect the library's strategy and its implementation, whereas the students are more likely to be affected by it. In this situation, we are also looking at the presence of a *dominant* stakeholder. The key point to stress is that the interests of the dominant stakeholder may not necessarily be the same as the interests of the other, majority stakeholders. This is perhaps particularly true of private sector organisations, where the 'profit motive' is likely to be dominant. It may nevertheless also be true of public sector organisations. Consider the interests of disabled users of technology in a library setting, for example. They are a minority group and, in the UK, until legislation was enacted, may as a result have been treated less favourably than other users. In such situations, there is often pressure on the dominant stakeholders to modify their self-interest to encompass the requirements of other groups.

Strategy is normally devised to take into account the requirements of all the stakeholders. 'The strategic management of technology is pursued under the influences and constraints of the objectives of the various stakeholders involved. Increasingly, there is a need to take into account a much wider range of stakeholders. The negotiation of objectives between different stakeholders is an important part of technology management' (Open University, 1994). Some of the stakeholders will have a direct input into the development of the strategy and/or be responsible for its implementation, monitoring and evaluation. Others will influence the strategy, but not be directly responsible for its contents. A third group will have a general interest in strategy and its implications. These groups can be categorised as in Table 3.2.

Analysing the stakeholders in an organisation, project, product, service or technology will take account of these three categories. Once this has been done, it is important to determine their requirements. These will naturally vary depending upon the

Table 3.2 Types of stakeholder group

Stakeholder group	Description
Primary	Direct interest in the strategy – typically employees, and especially senior managers and those groups and institutions that fund the programme of work that the strategy generates and that have a financial stake in the organisation to which the strategy pertains
Secondary	Indirect interest in the strategy – typically those who use the organisation whose strategy it is, or who supply goods and services to it
Tertiary	Remote interest in the strategy – typically those who are interested in the key actions of the organisation and how those actions are performed, such as government agencies or pressure groups

background and priorities of each stakeholder group. There are likely to be three themes that emerge from such an analysis. Firstly, stakeholders will have *demands* – things that they have to do and which they will need to have satisfied if they are to be supportive. Secondly, they will have *choices* and will wish to exercise them. Thirdly, they will be operating under *constraints* – such as a limited budget or an imperfect technical knowledge or current infrastructure. These will all need to be taken into account.

Technology assessment

Technology assessment (TA) is the long-term analysis of the impact and consequences of the technology being assessed. It is a technique that should be used regularly over the lifetime of a technology and of a technology strategy. TA can be problematic, especially where a technology is new and not yet proven. However, it is an important tool for ensuring success, because 'studies have shown that failed technology projects have only 10 to 15 percent of the effort devoted to analysis ... successful ones have 30 to 40 percent of the time spent on this front end work' (Andrews and Stalick, 1994). A TA normally consists of the stages shown in Figure 3.3.

Figure 3.3 Stages in a technology assessment

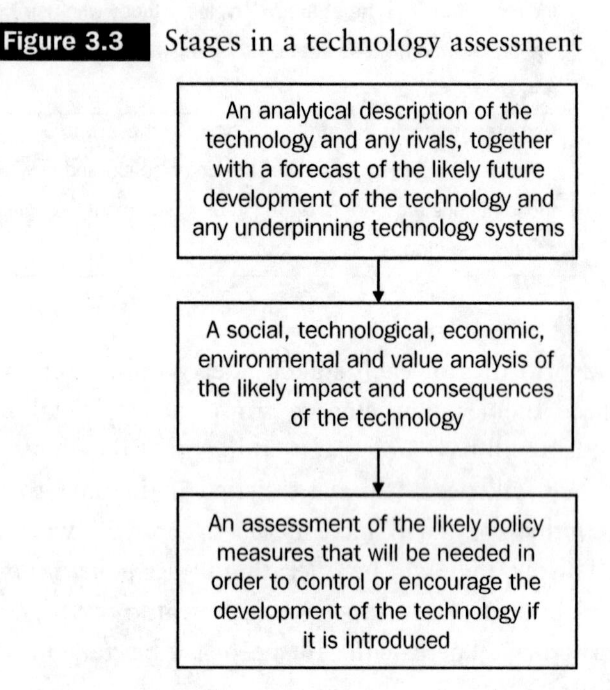

A good TA will need to consider opinion as much as hard data because current stakeholder views and future possible scenarios will need to be taken into account. A Delphi study might be useful, in order to have a clear view of both current and future environments in which the technology being analysed is likely to be used. The TA should also cover the impact on the organisation carrying out the TA. The impact is likely to be evident across a range of areas, such as training and development, resource allocation and management structure. In this context the TA might usefully be combined with the strategic/competitive web described earlier. Above all, the TA has to relate to the strategic aims and objectives of the organisation in order to provide 'opportunities to match changes in techniques, processes, and equipment to specific business goals and objectives' (White, 1988). Andrews and Stalick (1994) stress the need for an integrative approach in relation to IT, which 'plays a key role in any organisational change effort. Any change affects all parts of the organisation; therefore effort must address and integrate people, technology, structure and management philosophy.'

Organisational environment

In terms of the environment that is specific to the organisation's output, the *features* of the product or service need to be considered. These can be described as primary or secondary, depending upon their nature. Primary features concern the functions of the product or service; secondary features differentiate the product or service from others in the same field; tertiary features are those that the user discovers after first acquiring or using the product or service. In a document delivery service, for example, a primary feature might be the ability for end-users to order documents themselves; a secondary feature might be the facility to pay by credit card, unlike with some other systems; a tertiary feature could be the opportunity to integrate bibliographic searches carried out on the system with local citation software or catalogue services.

Grade encompasses the ability of a system to conform to specification and the 'tightness' of that specification. In a LIS unit, for example, a circulation desk might have a user waiting time of four minutes from the point at which the user joins a queue to the beginning of an issue or return transaction. The grade of service is a measure of the extent to which it is capable of achieving waiting times of under four minutes. Another LIS unit with a lower waiting time and the same level of performance would have a higher grade by virtue of a tighter specification.

Where technology is being acquired to enhance performance, the question of *availability* can be an important environmental factor. This relates to the time lag between the decision to acquire the technology and the point at which it is in use and making a difference. Some technology acquisition and implementation projects can be both time-consuming and of significant duration. An organisation's ability to provide rapid availability of new technologies can be an important factor in judging efficiency and effectiveness.

Price or *cost* is a key environmental factor. It includes the purchase cost, running costs and the eventual displacement cost.

Response time is important where an organisation is developing a new product or service. How quickly can it get a new idea to the point where it is a working and deliverable product or service? Sometimes, time will be of the essence, although this is arguably more an issue for industries that need to develop and market commercial products. None the less, LIS units that are slow to implement technologies that the general user can reasonably expect (e.g. self-service issue) to have available will be poorly judged in terms of their technology response time.

In addition to these features, the *general competence* of the organisation – the professionalism of its staff, its management, and its reputation based on past and present delivery – can also be a valuable environmental factor.

These features can all be combined in a competitive web that allows an organisation to compare itself with the 'industry standard' and also with key competitors or other benchmark organisations. Technology applications can play a part in improving performance in all the areas listed here. Cost reduction is one area where new technologies have often promised increased efficiency and productiveness, although they have not always delivered it.

Innovation scorecards

Innovation scorecards provide a way of assessing the extent to which an organisation is capable of innovating. They are a form of balanced scorecard that is useful for assessing the effectiveness and potential of an organisation overall (Stewart, 2001). The UK's Department of Trade and Industry (DTI)[2] produces helpful guidance and exemplars on these and other types of scorecard. The basic approach is to rate the organisation under a number of headings, such as in Table 3.3.

Table 3.3 Typical rating of an organisation

Rating	Level
1	No real level of innovation, and no plans to innovate
2	A basic level of innovation is evident, with a degree of motivation among the senior management to change
3	Innovation is a stated and implemented key objective of the organisation
4	The organisation is 'best of breed' in innovation and innovatory performance

This rating can then be used to score the organisation against key headings. These might be taken from the organisation's strategic plan, or be developed from the results of other analytical techniques such as a competitive or strategic web (see below). Once all the scores have been assigned, then they are put on the complete scorecard and the areas for improvement are identified and prioritised.

Gap analysis

What is more important in strategic technology management is not so much to identify existing gaps and fill them – important as that is for operational technology management – as to create new gaps that stretch the service, the sector, the market or all three. This is an essential part of innovation and continuous improvement in technology management. A gap might relate to competitiveness, where a new technology might be introduced, thus creating a gap between the existing dominant design and a new one around which innovation will now revolve. The organisation that fills the gap first will be able to lead the market. Sometimes there is a gap in the market that needs to be filled, though the technology may not be available to exploit that gap immediately. Or, the gap may come from an existing product using a prevalent technology not being distributed to a particular market. Case study 4 considers the Sudan, where there is clearly a market for mobile phone technology (the participants in the Khartoum workshop certainly all had one!). Sometimes, the gap comes from not recognising a possible usage for a technology. This may be particularly the case where two or more technologies are joined together in new ways. This is discussed in Case study 1, where the integration of Internet and LIS management systems has created a new kind of document delivery service. At the same time, commercial suppliers have recognised that there was a gap in the market and supplied such services from another kind of gap analysis.

Value chain analysis

A value chain analysis (VCA) is designed to allow organisations to identify gaps or weaknesses in the 'chain' of activities that are undertaken in order to produce their goods or services. Once identified, weaknesses can be reduced or eradicated in order to improve competitive advantage (Porter, 1985). A VCA may encompass more than one organisation; the key point to remember in carrying out the analysis is that only those activities under the control of the unit carrying out the VCA and able to make a difference to those operations requiring improvement should be included. Porter (1985) argues that because 'technology development consists of a range of activities that can be broadly grouped into efforts to improve the product and the process' (i.e. a value chain) then good technology management can be a powerful force in improving competitive advantage.

Value chains will vary in nature and composition depending upon the organisation, its alliances, markets and overall environment. However, it is generally recognised that there are certain basic elements common to all chains. These are described below. Note the similarities within the value chain to the input–transformation–output process discussed in Chapter 1.

The generic value chain consists of two parts. The upper or primary layer comprises five distinct blocks (see Figure 3.4). These are supported by four generic activities that the organisation must carry out in order to function fully.

Figure 3.4 Generic value chain

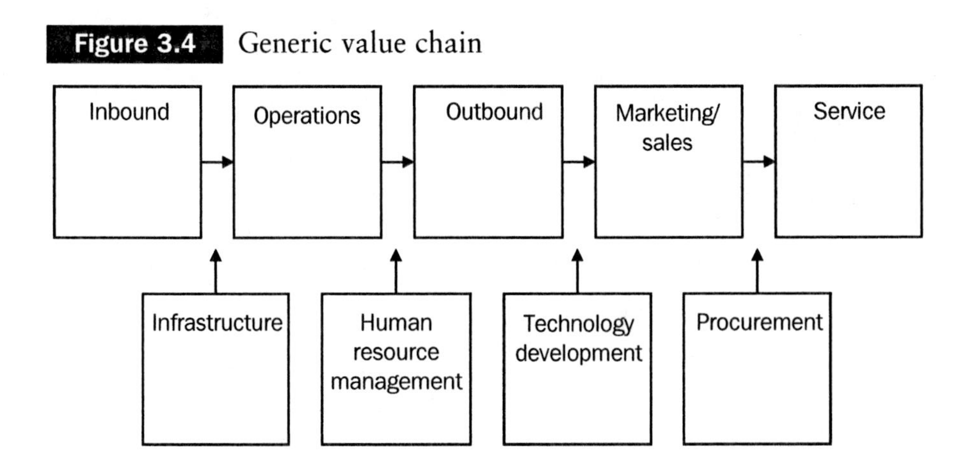

1. *Inbound* activities include everything relating to the acquisition and management of all necessary inputs to the organisation, for example hard copy or digital content and related cataloguing records or metadata.
2. *Operations* relate to those transformation activities that are the organisation's core business, such as the cataloguing and provision of library materials.
3. *Outbound* activities include everything associated with the distribution of the organisation's products or services to the buyers or users. These may link with operations in the case of LIS units, although specific circulation or Inter-Lending and Document Request Management System (ILDRMS) activities should be listed under this category.
4. *Marketing/sales* is that group of activities that attract buyers or users to the product or service or otherwise enable them to acquire or use what is offered. Most LIS units have some form of marketing or sales function, even if it is a not-for-profit organisation and the activity is designed to encourage usage rather than increase business in a strictly commercial sense.
5. *Service* in a commercial context might be labelled 'after-sales' or relate to maintenance of the product once acquired by the customer. In a LIS unit, this kind of service is more likely to comprise the reference and information or related help services that support usage of the services that, in user terms, make up the outbound element of the value chain.

In order to complete the VCA, each and every activity within the five primary and the four secondary activities is costed and its contribution to total revenue calculated. Where can the margins between cost and contribution be improved to increase competitive advantage? As Porter (1985) stresses, 'value activities are ... the discrete building blocks of competitive advantage. How each activity is performed combined with its economics will determine whether the firm is high or low cost relative to its competitors. How each value activity is performed will also determine its contribution to buyer needs and hence differentiation [from its competitors]'. VCA is often linked with business process re-engineering (BPR) discussed earlier.

SWOT analysis

SWOT stands for Strengths – Weaknesses – Opportunities – Threats. It is a technique that can be used to summate the results from the other analyses, such as gap and VCA. The following definitions are taken from Siess (2002)

SWOT analysis

A technique for analysing the environment in which the library operates or may operate in the future. It includes the following:

Strengths: areas in which the library has strong capabilities or a competitive advantage, or areas in which the library may develop capabilities and advantages in the time period covered by the strategic plan.

Weaknesses: areas in which the library is lacking the capabilities necessary to reach its goals, or areas that can be expected to develop within the time period covered by the strategic plan.

Opportunities: situations outside of the library that, if capitalised on, could improve the library's ability to fulfil its mission. These may exist now or develop within the time period covered by the strategic plan.

Threats: situations external to the library that exist now or may develop in the time period covered by the strategic plan that could damage the library and should be avoided, minimised or managed.

The analytical techniques described earlier in this chapter can all be used to provide the data that should inform a good quality SWOT analysis. The end result of the SWOT process should be that the organisation has determined either to move out of those areas where it is weak or to strengthen them in some way – as for example by creating a strategic alliance with another organisation that has the necessary strength.

The first stage will be to create a business profile on which the SWOT analysis can be based. The profiling process asks a number of questions about the nature of the business, the actual and potential customers, the ways in which the business is undertaken, the competition and the values that drive the business. Table 3.4 shows a worked example based on the EDDIS (Electronic Document Delivery: the Integrated Solution) project, described and discussed in Case study 3. Table 3.5 shows the SWOT analysis proper.

Table 3.4 EDDIS business profile

Nature of the business	To develop and sell a high-quality integrated document search, order, retrieve and management service to end users
Customer base	Academic libraries, primarily in the UK
Potential customer base	World wide
Customer requirements	A robust, cheap, easily managed system that will cut on-site journal and book holding costs while expanding end-user choice and maintaining central accounting and collection development control
Nature of business management	Via a project consortium, using PRINCE methodology, spread across the UK HE sector, with specialist software development partners
Why are we in business?	To deliver on an e-Lib project; to improve services and resource management in UK HE; to make money for the consortium
Competitors	Library system suppliers; publishers; other e-Lib projects; software development houses
Competitive assessment	Average to fair
Business values	To make the best use of taxpayers' money by increasing efficiency

In the SWOT analysis in Table 3.5, each attribute is marked out of 4, where 1 is low and 4 is high. The attributes may vary, depending upon the environment, the organisation or the technology. Having scored all the SWOT attributes, a SWOT grid can then be created, with the four elements S–W–O–T identified.

Table 3.5 **EDDIS: SWOT analysis**

Strengths and weaknesses			
Marketing			
A.	Appropriate product	3?	
B.	Marketing ability	1	(Commercial partner 3?)
C.	Correct promotion	1	(Commercial partners 3?)
D.	Correct price	1	
E.	Effective distribution	1	
Synergy and balance			
F.	Relations with partners	2?	
Management			
G.	Technical skills	2	(Commercial partner 4)
H.	Business competence	2	(Commercial partner 4)
Operations			
I.	Support	1	(Commercial partner 3)
J.	Development	1	(Commercial partner 4)
Finance			
K.	Capital	2	(Commercial partner 4)
L.	Recurrent	1	(Commercial partner 2)
Technical			
M.	Software	2	(Commercial partner 4)
Opportunities and threats			
Economic threats			
1.	UK HE	2	
2.	International markets	4	
Social and political factors			
3.	End-user predominance	4	
4.	Librarian support	1	
5.	Resource sharing	4	

Table 3.5 Continued

Products and technology		
6. Standards compliance	4	
7. Leading edge	4	
8. Widespread application	1	
Demographic factors		
9. Increased IT literacy	4	
10. Remote learning and research	4	
Markets and competition		
11. Proprietary products	1	
Technology		
12. Bandwidth	2	
13. Internet	3	

	Threats	**Opportunities**
	1. UKHE 4. Librarian support 8. Widespread application 11. Proprietary products 12. Bandwidth	2. International markets 3. End-user predominance 5. Resource sharing 6. Standards compliance 7. Leading edge 9. Increased IT literacy 10. Remote learning and research 13. Internet
Strengths A. Appropriate product	Emphasise the long-term value of EDDIS to the HE community	Build alliances with user groups in the community Build alliances with standards agencies and users Build alliances with resource-sharing consortia Build alliances with similar projects in other countries

Table 3.5 Continued

Weaknesses		
B. Marketing ability C. Correct promotion D. Correct price E. Effective distribution F. Relations with partners G. Technical skills H. Business competence I. Support J. Development K. Capital L. Recurrent M. Software	Get out of marketing EDDIS Strike a deal with the commercial partner Negotiate a price with the commercial supplier that fits the user community's aspirations and ability to pay	Act as agent for the commercial supplier within the UK HE community and perhaps elsewhere?

Competitive and strategic planning webs

The competitive web is a technique widely used in the private sector for benchmarking an organisation against its competitors. At its simplest, the technique is intended to facilitate discussion between managers with regard to the key indicators on which the best-performing companies in the sector are judged as being the best and against which the organisation undertaking the benchmarking exercise wishes to compare and contrast itself and its immediate rivals. In the context of not-for-profit organisations, like most LIS operations, a competitive web is more appropriately described as a strategic web. The principles are the same: the key headings under which the organisation is to be compared and contrasted with the rest of the sector and key 'rivals' are determined. Each heading is placed at the end of a line, the lines intersecting to form a star shape. The sector norms are then placed on the lines: the higher the performance, the further away from the centre and the nearer to the tip. The organisation's own performance and that of any other specific institutions is then also plotted on the lines to form a web from which it is visually possible to identify the main areas for improvement or high performance against the benchmarks. The creation of the web can be underpinned by as much or as little analysis as the participants in the process think fit. The point to stress is that for the exercise to be meaningful it must be based on sufficient 'hard' data to provide results that can be used to take key strategic decisions. The provision of such data is discussed later in this book, with particular reference to hard and soft systems methodologies.

The strategic planning web reproduced in Figure 3.5 was developed for the SCAITS (Staff Computing and IT Skills) project at the University of East Anglia. Each arm of the web represents a key element of the strategy being developed. It can be thought of as a continuum from 0 in the centre to 10 at the outer rim: 0 means zero progress; 10 means that the goal has been achieved. In each box is an agreed goal in relation to each key element. The point marked on each of the arms represents a judgement as to how far the organisation has progressed towards achieving its goal. Joining the points creates the web. The web can be made comparative by adding the equivalent positions for comparator or competitor organisations. This allows the strategy manager to see where the institution is placed relative to other organisations with which it benchmarks itself. The web is essentially a diagnostic tool. It does not claim to be a refined scientific instrument. What is important is the process of thinking through not only goals in relation to the key elements of the strategy, but also about what constitutes evidence of progress and where the major strategic and operational gaps actually are.

The next step is to judge the extent to which goals have been achieved and, most importantly, what still needs to be done. Having done this, the goals should be revisited and either redefined to make them more realistic aspirations and/or to think about what needs to be done to make progress along each

Figure 3.5 The SCAITS strategic planning web

Where are you now?

A strategic framework for C&IT skills
(from report section 4.1)

- Does your HEI have an information strategy linked to targets and resources?
- Does your HEI have a human resources strategy?
- Does your HEI have a strategy for learning and teaching?
- Does your HEI have a research strategy?
- Does your HEI have a C&IT skills strategy supported by a C&IT skills development policy?
- What is the linkage between the different strategic strands?
- Does a member of the senior management team have responsibility for infrastructure and C&IT skills development?
- Do departments have their own strategies for developing C&IT in learning and teaching?
- Is there evidence of collaboration between key support services and decision making bodies? (computing, audio-visual services, staff development, personnel, learning and teaching committee, information strategy steering committee, human resources policy committee, etc)?
- What are your funding arrangements and how do they influence strategic planning?
- How is your HEI organised? What impact does the decision-making structure have on the development of strategy?
- Is C&IT skills development a managed or an ad hoc process?

C&IT users – skills for all staff?
(from report section 4.2)

- Considerable variation is found between departments in terms of C&IT use by staff.
- Considerable variation is found between categories of staff in terms of C&IT use.
- All staff who wish to can become C&IT users.
- All staff are automatically given an e-mail address.
- All staff have access to appropriate C&IT skills training.
- There are pockets of non-users in all categories of staff.
- Take up of training courses is greatest amongst administrative, secretarial and clerical staff.

C&IT infrastructure and desktop provision
(from report section 4.3)

- Many staff have convenient access to desktop computers, but sharing is not uncommon. A significant minority of staff use desktop computers that run operating systems that are old and no longer actively supported within the institution. Networks and servers just about cope with normal loads but struggle at times of peak demand. Breaks in network based services of at least a half day duration happen several times a year.
- Most staff have convenient desktop computer access, and sharing is rare. Most run institutionally supported operating systems. Networks and servers cope with demand in all except the most unusual of circumstances. Breaks in network based services happen several times a year but normally last no longer than two hours.
- All staff have convenient desktop computer access, and sharing is extremely rare. With the occasional exception for demonstrably good reason, all computers run institutionally supported operating systems. Networks and servers cope with the load in all present circumstances, and breaks in networked based services are rare, with back up systems and network configurations enabling most breaks to be of no more than 15 minutes.

Appropriate C&IT skills for staff
(from report section 4.4)

- There is no agreed set of C&IT skills for any role in the institution.
- Some departments have identified C&IT skills requirements for at least some roles.
- A *basic* C&IT skills requirement for all staff in the institution has been identified.
- Besides the basic skills requirement, additional C&IT skills have been identified for many roles.
- All roles have an associated C&IT skills profile.

C&IT training and development
(from report section 4.5)

- Some C&IT training is offered by departments, with some central provision. Attendance is entirely by individual choice.
- Major central provision is occasionally supplemented by departments. Attendance is largely on the basis of individual choice.
- Major central training provision, though still within a voluntary framework, is supplemented by departments, with some departmental focussed training.
- IT training is delivered in a variety of ways: central course provision, departmental focussed training, learning resources centre, one-to-one workplace coaching, self-instructional packages, problem based worksheets.
- C&IT awareness training is provided for all managers.
- Emphasis is on recruiting staff with the requisite skills. Access to training is determined by line managers and linked to specific job requirements.
- Regular C&IT training needs surveys are linked to C&IT skills development policy and C&IT skills strategy.

Help and support when using C&IT
(from report section 4.6)

- Central Help Desk is usually provided for part of the normal working day. Some departmental technical support is available.
- Central Help Desk provided *throughout* the normal working day (including lunch time).
- Central Help Desk is supplemented by departmental support staff (or replaced by Departmental Help Desks). Self-help leaflets and/or on-line FAQs available.
- Readily available formal help as above. Also electronic discussion groups/lists for mutual help among users, and system of designated local support for each office and work group.

C&IT qualifications and professional development
(from report section 4.7)

- No opportunities are offered by the institution for staff to gain C&IT accreditation or qualifications, although C&IT training is available.
- As above but staff also have access to externally run accreditation.
- Some C&IT accreditation is offered in-house, but take-up largely governed by individual choice.
- A basic C&IT accreditation system is available in-house. Staff are either recruited with the requisite skills or encouraged to demonstrate competence through accredited programmes.
- C&IT job related skills are specifically linked to levels of accreditation. Staff are either recruited with appropriate qualifications or required to undertake training within specified timescales.
- Staff are given financial support to gain specialist C&IT qualifications.

C&IT skills and recruitment
(from report section 4.8)

- There is no institutional requirement to produce job descriptions or person specifications. Advertisements may or may not indicate levels of C&IT skills. There is no monitoring of the process.
- Skills are assumed but untested at interview. It is the responsibility of the individual to learn the skills necessary to do the job. The profile of C&IT skills in recruitment is variable.
- All posts are linked to job descriptions or person specifications. All recruitment literature includes details of required and desirable skills. The recruitment process is monitored centrally.
- All applicants are required to provide evidence and/or be able to demonstrate at interview a level of C&IT skill appropriate to the post.

C&IT skills and promotion
(from report section 4.9)

- C&IT skills are rarely explicitly considered in the promotion process, other than for technical IT related posts.
- Criteria for promotion include some reference to C&IT skills.
- Appropriate levels of C&IT skills are necessary before staff can be promoted.
- The demonstration of exceptional levels of job-related C&IT skills is a criterion for promotion.

Where do you want to be?

Figure 3.5 Continued

The SCAITS Strategic Planning Web

Creating a strategic planning web for your institution

For an example, see Section 3 of the report.

To draw your own web:

1. Read through Section 4 of the report, which is divided into nine parts, each focussing on a key element. At the beginning of each part there are statements (repeated on the back of this blank planning web) which will help you position your HEI and help you make a judgement as to where you are in relation to your goal.
2. On this proforma each of the nine arms of the web is associated with one of the key elements. State your provisional goal for each element in turn, in the appropriate blank box.
3. Each arm can be thought to represent a continuum, from 0 in the centre to 10 at the outer rim of the web. 0 means zero progress, 10 means the goal has been achieved.

   Make your judgement based on the notional scale of 0-10 as to where your HEI is in terms of achieving each goal, and mark each arm accordingly.
4. Join up the points to create a web.
5. Think about what the whole picture says concerning staff C&IT skills in your HEI. Do the goals complement one another? Is there one element that is having a disproportionate impact on the others?
6. Identify future action and/or revisit your goals.

This mapping exercise can be done individually but it has much more value if used as part of a planning process. The final version then becomes the agreed outcome of a longer consultation process between different interests.

Supplied with the JCALT report 'C&IT skills: Developing staff C&IT capability in Higher Education'

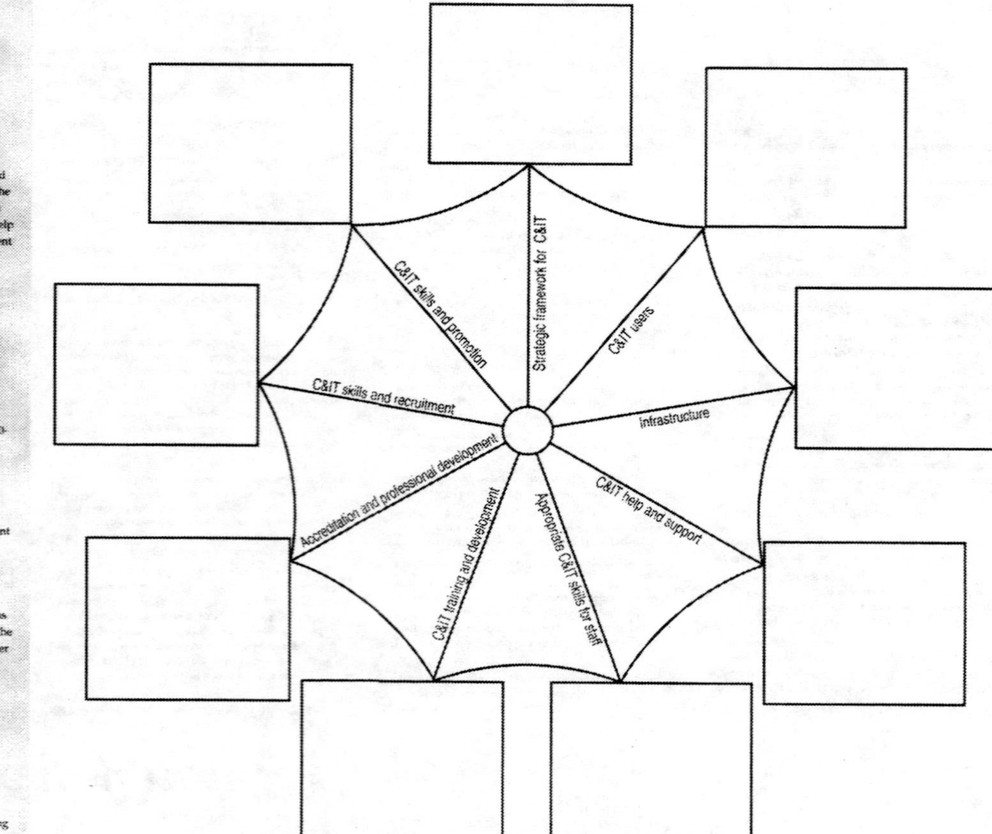

Reproduced courtesy of the Joint Information Systems Committee and the University of East Anglia.

continuum. There may be some actions, such as a decision to undertake an organisation-wide survey of ICT skills, that will provide the basis for action in a number of areas. Some actions will have a wide impact; others need to occur in parallel and do not form a simple sequence.

In an ideal world, the web drawn by a middle manager will look the same as one drawn by the chief executive. This is another way of taking the 'Strategic IQ test' referred to in Chapter 1. If, in practice, they diverge significantly, then there is an issue to be addressed as part of the process of creating the overarching strategy. For example, if the goals suggested by individual departments demand a high degree of autonomy in relation to the purchase of hardware and software, high levels of support and a next generation infrastructure, then in the presence of resource constraints this may not be consistent with, say, an institutional goal of raising the levels of basic ICT skills for all staff.

Summary

This chapter has considered strategic and technology foresight and forecasting in the context of present and future environments and organisational position. It has looked at the major techniques that can be employed. The Delphi method provides a future framework or set of scenarios in which a number of more specific analyses can be carried out: reviewing the organisation's portfolio of products or services is one; another is the market in which that portfolio will have to be offered. These analyses can usefully contribute to a product development mapping exercise, where enhancements or other changes can be considered and prioritised. Quality function deployment is a technique for matching customer requirements with product or service development.

Customers or users are only one of the stakeholders, and there will be a need to analyse all the stakeholders' needs and priorities.

This is especially important when developing strategy. Where technology is an enabling factor (rather than the subject of the development) there will be a need to assess that technology in terms of its long-term future. This assessment also needs to look at the organisational environment and the needs that the technology is meant to satisfy.

Reference was made in Chapter 2 to an organisation's ability to innovate and innovation scorecards can be used to determine what the real capacity is. This exercise may reveal gaps in the organisation. A gap analysis is concerned not just with existing gaps but also the anticipation of future gaps that the organisation could effectively fill. A value chain analysis is another way of looking at this, with special reference to the product, the process and the difference that improvements to elements in the chain will make to competitiveness.

SWOT analyses are tried and tested, and still have a significant value, especially in strategic technology management. An important prerequisite will be the creation of a business profile that states the main aims of the organisation.

The chapter ends with a proposed strategic planning technique based on a web of goals that need to be taken into account. Creating the framework for the web makes the strategic technology manager think about what the basic elements of the strategy are and the assessment tools can provide data with which to draw the web itself, including the position of other benchmark organisations and/or an ideal benchmark for the sector.

Notes

1. *http://www.ingenta.com/*
2. *http://www.dti.gov.uk*

Bibliography

Andrews, D.C. and Stalick, S.K. (1994) *Business Reengingeering: The Survival Guide*. New York: Prentice Hall.

Fahey, L. and Narayan, V.K. (1986) *Macroenvironmental Analysis for Strategic Management*. St Paul, MN: West Publishing.

Feather, J. (2003) 'Theoretical perspectives on the information society', in S. Hornby and Z. Clarke (eds), *Challenge and Change in the Information Society*. London: Facet, 3–17.

Hamel, G. and Prahalad, C.K. (1994) *Competing for the Future*. Boston: Harvard Business School.

Kodama, F. (1991) *Analyzing Japanese High Technologies*. London: Pinter.

Kotler, P. (1988) *Marketing Management*. Englewood Cliffs, NJ: Prentice Hall.

Open University (1994) *T841: The Strategic Management of Technology*. Milton Keynes: Open University.

Porter, M.E. (1985) *Competitive Advantage: Creating and Maintaining Superior Performance*. London: Collier Macmillan.

Siess, J.A. (2002) *Time Management, Planning and Prioritisation for Librarians*. Lanham, MD: Scarecrow.

Stewart, W.E. (2001) 'Balanced scorecard for projects', *Project Management Journal*, 32(1), 38–54.

Twiss, B. and Goodridge, M. (1989) *Managing Technology For Competitive Advantage*. London: Pitman.

White, B.L. (1988) *The Technology Assessment Process: A Strategic Framework for Managing Technical Innovation*. New York: Quorum.

4

Strategy formulation II: systems thinking

Introduction

In order to formulate and then implement strategy effectively, it is necessary to have a full understanding of the situation to which the strategy pertains. Without this understanding, currently held assumptions may prevail when they need to be challenged and, indeed, overturned. This is especially true of strategic technology management, where too often assumptions are not tested sufficiently well and major mistakes are subsequently made as a result. It is especially tempting to think that enough is known about a situation before embarking on implementation through specific programmes of work and individual projects. LIS staff are as much under pressure as any group of public or private sector workers. Time is money and having a competitive edge is a finite state. However, 'more haste, less speed' remains a crucial adage to remember.

This chapter therefore encourages the technology manager to take as objective an approach as possible to strategy formulation and its subsequent implementation. Using a systems approach, as described later in the chapter, will help to ensure that a holistic view of the strategy, its environment and its implementation pathway is adopted. Because this book is about strategic management, it does not cover those techniques that are useful for operational change and incremental improvement – the normal, everyday management of quality improvement – but rather the innovatory levels of change that require the higher-level strategic application techniques to be managed effectively and efficiently. It is based upon the idea of the *model*: a set of organised assumptions about a particular aspect of the world and the way that it works. Modelling allows managers to select from, and simplify, all the possible information that is available relating to a given problem or opportunity. The more complex the model, the more planning techniques will be required in order not only to incorporate all the required data, but to organise it in a meaningful way. The model is most useful in predicting the way that a system will behave in given circumstances, and especially when a particular route towards set objectives is being proposed.

Strategic and operational levels

It is important to focus down to the operational level when considering strategy implementation. In doing this, it is assumed that the strategy itself has been properly formulated and is ready to be implemented, and that the 'bigger picture' is already well known and understood. The emphasis in strategy implementation is thus on the operationalisation of the aims and objectives set out in the strategy. This requires the systematic reduction of the problems implicit in the strategy as formulated. Two approaches – broadly labelled as reductionist and systems-based – are therefore considered. Although at the operations level managers are more concerned with present-day issues, it is nevertheless of crucial importance to ensure that the approaches are well integrated with the longer-term view. This is vital in the case of technology strategies, where so much can change so quickly and implementation pathways that were appropriate when the strategy was first introduced become obsolete.

As discussed in Chapter 1, a strategy has a number of aims and objectives. The objectives form the basis of an action plan or programme of work that actually results in 'things getting done' if carried out effectively. This will require some kind of transformation process that takes the organisation from 'where it is now' to 'where it wants to be in the future'.

What information is required in order to formulate and then implement strategy effectively and efficiently?

It is of fundamental importance that decisions are taken on the basis of adequate and appropriate information. Even at an operational level, it is crucial to put the application of technology into a broader context. As the previous chapter demonstrated, there are many ways in which a technology strategy and its implementation can be 'blown off course'. A considerable number of these are related not to technology but the overall environment in which the technology is being introduced or re-engineered. Many of the strategic planning techniques described earlier in this book can also be used to check the viability of the strategy in operational terms. Two such techniques are the 'strategic IQ test' referred to in Chapter 1 and the strategic web discussed in Chapter 3. Between them, these approaches provide the technology manager with a useful summary of the position with regard to the organisation and its readiness for technology change, as set out in the strategy. In particular, the human resource aspects of the operations are covered. As Anstey (2000) points out, without an appropriately skilled workforce and a technology provision that is fit for purpose, any technology strategy will fail. People are the fundamental resource of any project and in all organisations and the effects of the technology on staff and vice versa need to be well known both at the strategic and the implementation levels.

What types of thinking are involved in order to implement strategy?

Reductionist thinking assumes that there is a single, structured problem with a single desired outcome. In order to use a reductionist approach effectively, larger problems have to be broken down into a series of single problems. Reductionist thinking encourages the technology manager to disassemble the objective into its constituent parts, find out the key problem area, provide a solution and then reassemble the objective in the expectation that the changes to the particular area identified and altered will not affect the objective overall.

If combining the solutions to these single problems is unlikely to give an effective answer to the total problem, then systems thinking will be required. In systems thinking the manager has to take an overall viewpoint in which the whole is greater than the sum of the parts, not least because due account needs to be taken of the interactions between the major variables. In the case of strategy formulation and even implementation, it is highly unlikely that management will have control over any of the major variables and their interactions.

Reductionist thinking stresses efficiency; systems thinking is more concerned with effectiveness. Efficiency relates to how well something is carried out; efficiency to how appropriately it is undertaken. The two concepts are intimately connected: it may increase efficiency to reduce the hours during which a full guaranteed network service is provided, but the effectiveness will be reduced in terms of the quality of service offered to the end users and their likely rate of satisfaction with what is available, especially if the quality level falls short of that stated in the organisation's strategy. Table 4.1 sums up the main differences between reductionist and systems thinking.

Table 4.1 Reductionist and systems thinking

Reductionist	Systems
Problems are separated into simpler ones before a solution is attempted	A holistic view of the problem – in all its aspects – is taken
Results can be obtained quickly	More time is needed, at least to begin with
The sum of the parts is the same as the whole	The whole is greater than the sum of the parts
There is only one cause and one associated effect	There are many causes and potentially several effects
One small area only is normally considered	A wide area is normally considered
The situation is controllable	The situation is not completely controllable
The problem or opportunity to be analysed is well defined	The problem or opportunity to be analysed is not always well defined

The narrower the approach, the more likely it is that strategy planners and managers will fall into the trap of thinking that they know how to solve the challenges and problems associated with implementation on the basis of past experience without looking more broadly at the issues that really need to be taken into account before proceeding. The emphasis in much strategy implementation, then, is on holistic thinking. A library service is a complex system. In order to understand it properly, it is not sufficient to think of it as a place, say, to borrow books. It is a complete environment with a wide range of interactions between technologies and people, within both a physical and now a virtual space. Different people will have different perspectives on the system: a library user will be looking at the system from a different viewpoint from the library manager. Indeed, different users will have a wide range of varying perspectives: is the library a place to borrow books, check e-mails, do group work or have a coffee? Is it a physical space at all? Research staff in universities are increasingly proud of the fact that they can use the library very effectively over the Internet without ever going into the building!

Problem definition

Problem definition is a key aspect of the development and implementation of strategy through systems thinking. Typically, there will be a number of problems to be solved as part of the strategic technology management process. These will present a set of issues that the organisation needs to tackle as part of its strategic development. The problems are likely to spring from use of the various analytical techniques described in earlier chapters. Problems will vary depending upon the extent to which they, and their possible solutions, are already defined or are capable of being defined, and the degree to which the processes for investigating the problems and identifying and implementing solutions are defined or are capable of being defined. In other words, the degree of uncertainty will affect the ways in which problems are approached.

A number of techniques for developing and defining problems can be employed (Van Gundy, 1988). Three have been selected as being particularly useful in the context of strategic technology management in a LIS environment. They are: boundary examination, 5Ws and H, and dimensional analysis.

Boundary examination

The aim here is to define broad boundaries and measure the dimensions within which the problem can be contained but not constricted. This will, it is hoped, make the problem more manageable. In defining the limits of the problem, it is important not to be too rigid, as this may pre-judge the likely outcomes of what is meant to be a process through which the initial problem is redefined in order to open up new ways of looking at it and hence to provide solutions that will improve matters. The basic technique is highlighted in the example shown in Figure 4.1. An initial statement of the problem is created, from which the major assumptions are drawn. The key implications of these assumptions are then identified and a new problem definition or definitions is created in the light of these. The new definition can then be used as the basis for further work.

5Ws and H

This is another popular technique for defining and redefining problems. The aim is to gather information systematically so that the problem can be structured and, hence, properly defined. Although data collection is the prime purpose, the process can generate new perspectives and lead to a redefinition of the original problem.

The process begins with a statement of the problem, turning it into a question beginning with the words: 'in what ways might (IWWM) ...? Having stated the problem, a series of questions are asked under the basic headings of Who? What? Where? When? Why? and How? Once this has been done, the answers are examined and problem redefinitions suggested. These redefinitions might themselves be produced by turning the answers to the 5Ws and H questions into further questions beginning with the IWWM phase. The redefinition that most suits the answers to the questions is then selected.

Van Gundy (1988) recommends that 'double-barrel' problem statements should be avoided but that duplication of questions between different headings is unavoidable and, indeed, to be

Figure 4.1 **Boundary examination**

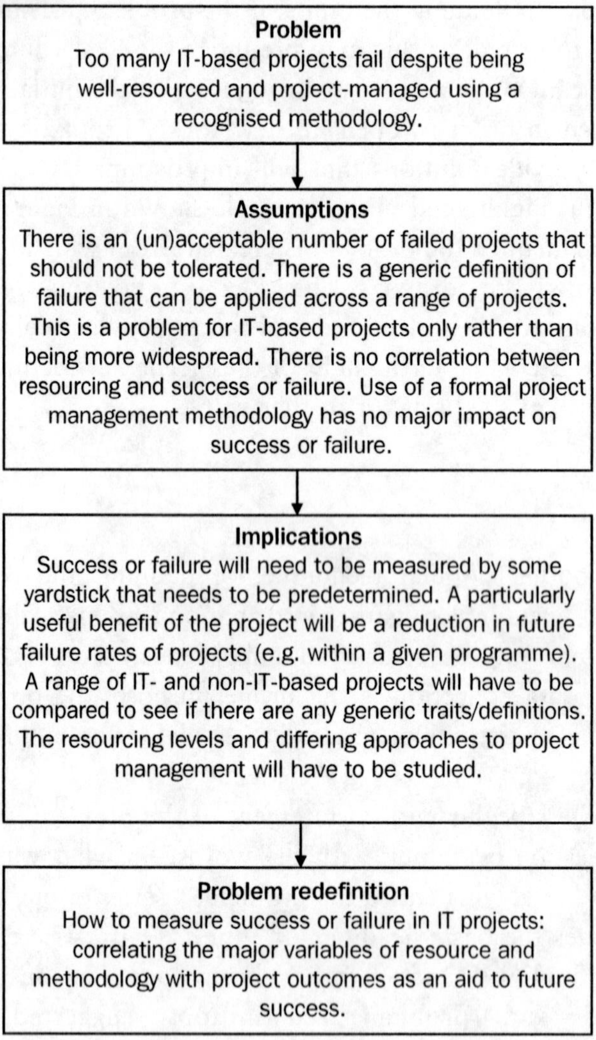

Problem
Too many IT-based projects fail despite being well-resourced and project-managed using a recognised methodology.

↓

Assumptions
There is an (un)acceptable number of failed projects that should not be tolerated. There is a generic definition of failure that can be applied across a range of projects. This is a problem for IT-based projects only rather than being more widespread. There is no correlation between resourcing and success or failure. Use of a formal project management methodology has no major impact on success or failure.

↓

Implications
Success or failure will need to be measured by some yardstick that needs to be predetermined. A particularly useful benefit of the project will be a reduction in future failure rates of projects (e.g. within a given programme). A range of IT- and non-IT-based projects will have to be compared to see if there are any generic traits/definitions. The resourcing levels and differing approaches to project management will have to be studied.

↓

Problem redefinition
How to measure success or failure in IT projects: correlating the major variables of resource and methodology with project outcomes as an aid to future success.

encouraged. The purpose of the technique is to draw out all useful information as appropriate. There is no single correct answer: the value of any problem redefinition will vary depending upon the context in which the problem is being identified and solved.

A worked example is given in Figure 4.2.

Figure 4.2 Five Ws and H

Problem

In what ways might success be better measured in IT projects through correlating the major variables of resource and methodology with project outcomes?

Who?

- ...are the key players in any project?

They are the people who can affect a project for better or for worse. They stand to gain or lose by the success or failure of a project. They are the people who commission, manage, supply or otherwise interact with the project in a way that ensures that the project either succeeds or fails in its original aims. They may be members of the project board, the project team or external suppliers (e.g. of software). The relationships to the project may or may not be of a contractual nature.

What?

- ...is success?

Success is the completion of a project on time, to budget and to agreed quality. Success is the delivery of what was promised, expected and hoped for. Success is the delivery of a product or service that does what the project instigators said it would do and which yields the promised benefits to those who subsequently use it and at whom it was aimed.

Where?

- ...can project success be measured?

See the answer to the question 'What is success?'

- ...do projects tend to go wrong?

In the time budgeting. In the management of rapidly changing technology. In the management of the people. In the setting and realising of aims and objectives.

When?

- ...do projects succeed?

When the project aims are realistic and manageable within the resource and other constraints (e.g. technology). When all the required project management competences are covered within the project team.

Why?

- ...do projects fail?

Because project aims are unrealistic and unmanageable within the constraints. Because not all the required competences are covered. See also the answer to the question 'When'.

How?

- ...can project success be assured?

By ensuring that the project aims are realistic and manageable within the resource and other constraints and that all the required project management competences are covered within the project team.

Problem redefinition

Predetermining success: analysing key project variables in order to ensure effective and efficient delivery of promised and projected aims and objectives.

Dimensional analysis

This method aims to identify the *dimensions* (and hence the limits or boundaries of a problem) under five headings:

- Substantitive
- Spatial
- Temporal
- Quantitative
- Qualitative.

The dimensions are defined by asking What? Where? When? How Much? How serious? This is followed by a more detailed analysis of the dimensions, as noted in Table 4.2. Table 4.3 gives a worked example, based on the EDDIS project described in Case study 3.

It should be noted that not all questions will be appropriate in all circumstances, and for library-specific problem analysis, the more philosophical dimensions may not be necessary. The main point to remember is that the technique is only a means to an end: the effective identification of a problem that needs to be solved as part of a broader strategy development and implementation process.

Table 4.2 Dimensional analysis

Substantive	Spatial	Temporal	Quantitative	Qualitative
Commission or omission? Does an activity need to stopped or a new one be started?	*Local or distant?* What geographical area is covered by the problem definition? Case study 1 discusses a problem that is international; Case Study 2 is at least national if not international; Case Study 3 concerns a national-level programme and a specific project that had implications beyond the partner libraries; Case Study 4 is linked to a particular country.	*Long-standing or recent?* The tension between making library stock accessible to users and secure from theft is a long-standing problem. The question of Value-Added Tax on electronic journals and not on hard-copy equivalents is a recent issue.	*Singular or multiple?* One or more factors may contribute to a problem. As Heeks et al. (1999) found, the effectiveness of technology application in the UK's National Health Service was undermined by a combination of demonstrations of 'technocratic utopianism' on the part of the developers and innate cultural conservatism among the workforce. It is suggested that singular item problems will be rare in technology management.	*Philosophical or surface?* Will it be necessary to question any underlying philosophical assumptions when considering the problem? The inadequacy of an individual library's acquisitions budget is a surface problem; the question of whether digital resources should be free to all at the point of use is a more philosophical issue.

Table 4.2 continued

Substantive	Spatial	Temporal	Quantitative	Qualitative
Attitude or deed? Does the problem originate from attitudes or behaviours? Changing attitudes can lead to changes in behaviour.	*Particular locations within a location?* The problem may emanate from, and be centred upon, a specific area. In a 'hybrid' library context, for example, it may be the traditional elements of the service that are insufficiently integrated with the new, ICT-based aspects that are causing the problem, rather than the service overall.	*Present or impending?* Is the problem already evident or will it only appear if current trends persist? If the latter, then there is an opportunity to prevent the problem from occurring unless it is beyond the power of those engaged in the strategy development process, in which case there has to be an appropriate response to the impending problem. The question of interoperability is a problem that is already present, but it is also an impending one in terms of new ICT applications such as managed learning environments (MLEs).	*Many or few people?* The cause or effect may relate to only a small group of people or to a whole population. In an academic library context, for example, a technology problem may relate only to a systems development group or to the whole campus. The problems of the one may impact on the other, of course.	*Survival or enrichment?* A problem of survival action, as with for example a significant overspend on an institution's budget such that remedial action will be necessary in order to avoid bankruptcy. The need to enrich a service is a less immediate or urgent goal, and one in which time can be taken to plan ahead.

Ends or means? Or what is the cause and what is the effect? It is perhaps easier to identify the symptom rather than the cause. An unwillingness on the part of academic staff to adopt ICT for teaching and learning (as discussed in Case study 2) may be a behavioural issue, but what are the underlying attitudes that need to be changed in order for a more desired behaviour to occur?	*Isolated or widespread?* Is the problem a 'one-off' or does it pervade the whole system? Is it restricted to one LIS unit or is it generic to the sector or subsector as a whole? A library might have an obsolescent ILDRMS. That is an isolated problem. An inability of suppliers to the sector to provide latest generation ILDRMS systems at affordable prices would be a widespread problem.	*Constant or ebb-and-flow?* Life cycles, such as those experienced in technology management, may create high and low points in the problem area over time. It may not be possible to predict when the peak of the cycle will be with any degree of accuracy; the more unexpected the problem, the more difficult it will be to respond to it effectively. Journal price inflation seems to be a constant and enduring problem. Some technology applications seem to be cyclical – as with for example developments in IT-assisted teaching and learning (ITATL), although the problem of take-up seems to be an enduring one, as noted in Case study 2.	*General or specific?* In other words, where is the most applicable area of the problem? Some forms of IT application will be of particular relevance to one part of the LIS sector, as for example special libraries and one-person bands.	*Primary or secondary?* This is a matter of priority, and perceptions may vary among different stakeholders depending upon their own interests, view or contexts. A primary problem for a library might be the urgent replacement of an obsolete system; a secondary one the level of provision of plug-and-go data points within the library building for users, though the users may see this differently!

Table 4.2 continued

Substantive	Spatial	Temporal	Quantitative	Qualitative
Active or passive? This relates to the threat posed by the problem. Is it one in which an active response is required in order to reduce the threat, or is the problem one with which the institution can live, even as an irritant, until such time as it needs to be tackled or goes away 'of its own accord'? In a competitive environment, a problem that posed a real threat to the future viability of the service, for example a traditional hard-copy document delivery process being threatened by direct end-user ordering and receipt over the Internet, would require an active response. So too would journal price inflation that crippled a library's ability to stock a			*Simple or complex?* Intellectual property right (IPR) and the Internet is a complex issue simply because it involves so many different stakeholders or pressure groups and is international in scope. As noted above, simple problems are likely to be rarer than complex ones in technology management, especially where future trends are not an obvious predictor of possible problems.	*What values are being violated?* This question is about determining what is wrong as opposed to what the problem is. Heeks et al. (1999) found that the real problem with ICT applications in the UK's National Health Service was the lack of communication between the various parties and, indeed, a high degree of arrogance on the part of system developers who 'knew what was best' for the users without actually asking them.

sufficiently appropriate range of titles. A passive problem might be the level of take-up of ICT-based tools among certain user groups – irritating but not perceived as threatening to the survival of the service, at least in the short to medium term.				
Visible or invisible? Is the real problem actually evident, or are we seeing the 'tip of the iceberg'? This is particularly problematic with human issues in relation to the take-up of technology, as Heeks et al. (1999) discuss.			*Affluence or scarcity?* Supply and demand can cause problems. If there are too many firms competing in, say, the library management market, then corners may be cut inappropriately in order to gain business. The customer then suffers after placing the order through poor service caused by cost-cutting on software maintenance. Conversely, if a new and highly desirable product is only available from a small number of suppliers, then	*To what degree are values being violated?* In other words, how serious or how trivial is the 'violation'. A lack of consultation between a systems team and the library supervisors may be trivial in the context of scheduled downtime about which staff members were not informed. The example given by Heeks et al. (1999) as noted above is much more serious.

Table 4.2 continued

Substantive	Spatial	Temporal	Quantitative	Qualitative
			the price may well increase, or the successful suppliers may be unable to cope with demand and again service to signed-up customers deteriorates.	
				Proper or improper values? How important and inviolable is the value that has been identified? Free access to digital resources at point of use may be regarded as a high value, although it may not always be possible – thanks to budgetary constraints – to honour that value. In an increasingly market-orientated public sector, it could be argued that it is no longer an important or proper value.

Table 4.3 Dimensional analysis: a worked example

Dimension	Comment
Substantive	The problem includes 'sins' of both commission and omission (see the above analysis). Project problems stem from both attitudes (e.g. to new technology) and deeds (project mismanagement). The analyses above also suggest that determining cause and effect in project failures will not be easy. There are likely to be two sets of problems: visible system-type problems (e.g. software bugs) and invisible people-type problems (e.g. antagonisms between software designers).
Spatial	There is no particular location to the problem. According to the literature search results, it has many generic elements. However, the particular study's concentration on one set of projects (the e-Lib programme) and the review of one particular project in particular (EDDIS) provide spatial boundaries for the problem: the UK higher education sector and the partners in the e-Lib programme and the EDDIS project.
Temporal	The literature search strongly suggests that the problem is a long-standing one, although the emphasis on technology is a relatively recent added variable which can increase the cost of failure (e.g. the TAURUS and ARAMIS projects). The problem has existed for some time, therefore, and recent literature suggests that it is likely to continue into the future, despite the wealth of 'how-to-do-it-good' texts that now exist.
Quantitative	The problem exists for multiple reasons as noted in the above analyses. Not only are there issues to do with identifying which variables are causing the problem in given circumstances, but there is also the added difficulty of the *interaction* between the variables. In the case of EDDIS, for example, was the difficulty that of working with an innovative but ill-disciplined software house or the untried nature of the technology or the constraints put on development by the e-Lib programme board? Or all three?
Qualitative	The question of project failure is a fundamental one, as evinced by the significant literature on the subject. It is more than just a simple practical problem. It has deeper systemic and philosophical elements; it has a soft as well as a hard edge to it. The problem is one of how to enrich the ways in projects can be undertaken successfully. However, because 'the best is the enemy of the good', there are many who may say that project management is 'good enough' (this is certainly the flavour of the e-Lib programme evaluation reports). Is there any violation of values in project failure? It could be argued that misuse of taxpayers' money on projects that do not yield tangible and practical results violates the value of good citizenship on the part of those who spend public money on innovation programmes such as e-Lib. However, negative results can be as useful as positive ones and there is a need to explore the extent to which perceived project failure actually can contribute to programme success in the longer term.

Conceptual problems

So far, the discussion has concentrated on 'real-world' problems that require specific solutions, or at least a range of realisable options for likely success. However, as is evident from certain questions in the dimensional analysis as described and exemplified above, there are often important underlying conceptual problems that emerge and, indeed, that need to be identified. This is especially important in those circumstances where important lessons need to be learnt for the future, in order both to avoid further mistakes being made and to ensure that strategies can meaningfully respond more effectively to future challenges.

One way of identifying underlying conceptual problems is to categorise them into broad *areas*. 'IT project management' might be one such broad area. Within the given area, a number of *fields* might then be designated, for example 'the management of innovation'. There will then be a number of *aspects* to a field. Within innovation management, for example, there will be a number of aspects, such as the management of specific technologies, or the markets in which the innovation may be introduced and then there will be the management of the risk that will inevitably occur where an innovation is being launched in a new market. This classification is a simple one and creates a hierarchy between the different concepts that may emerge from an analysis of identified problems in strategic technology management. However, there is a need to group the areas, fields and aspects into a set of relevancies. A *relevance tree* is a form of diagramming that provides the strategist with a way of linking the areas and their subconcepts in a meaningful way. The results can then be used to draw generic conclusions from specific problem analyses.

Figure 4.3 shows a worked example of a relevance tree that draws on the earlier exemplars that highlighted the various definitional techniques.

Figure 4.3 Relevance tree

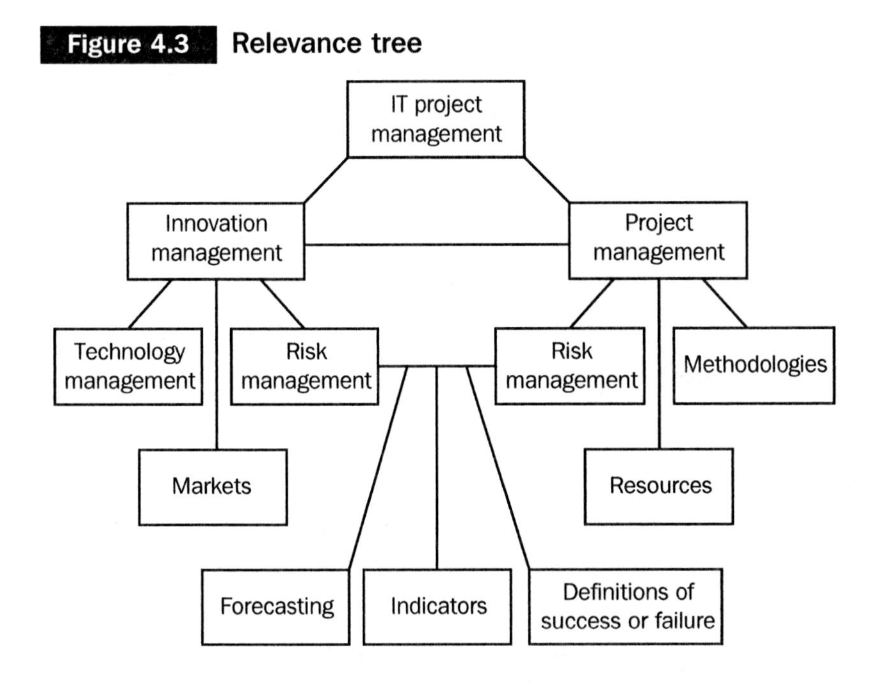

Systems thinking

A system can be defined as a set of components interlinked in an organised way for a specific purpose, for example a transformation process. The components within this assembly interact with each other in a particular way; changing or removing one of the components from the assembly will alter the way in which the system works. The interactions between the various components may not always be easy to identify. In human interactions, for example, unofficial channels of communication complement – or even contradict – official ones. 'Corridor chat' is one form of communication that is often overlooked in systems implementation and the management of change. Changing the official channels without taking account of the unofficial ones may mean that the new system is ineffective.

Feedback systems

The interactions that control the system are particularly important because, if changed, they are the most likely to control the behaviour of the system. These interactions are known as feedback control mechanisms. There are two types: feed forward or open loop control, and feedback or closed loop control.

Feed forward systems

A feed forward system is based on given and predictable knowledge (based on past experience) of a system, its inputs and its outputs (Figure 4.4). The system is controlled through the input of information that will control the process once it is started. The resulting control plan governs the process until it is finished or there is some intervention outside the system. Feed forward systems are typically used where there are predetermined outputs resulting from processes that always yield the same results. An Inter-Lending and Document Request Management System (ILDRMS) that is linked to the British Library's Document Supply Centre could be described as one such system. The supply process can only behave in a set number of ways to provide predictable outputs – all of which will be well known to the document supply librarian managing the process at the requesting or input end of the system. A feed forward system works well in these circumstances but cannot cope with the unexpected. What if the responding library does not provide output in the expected format? Some form of feedback process is required.

Figure 4.4 **Feed forward control**

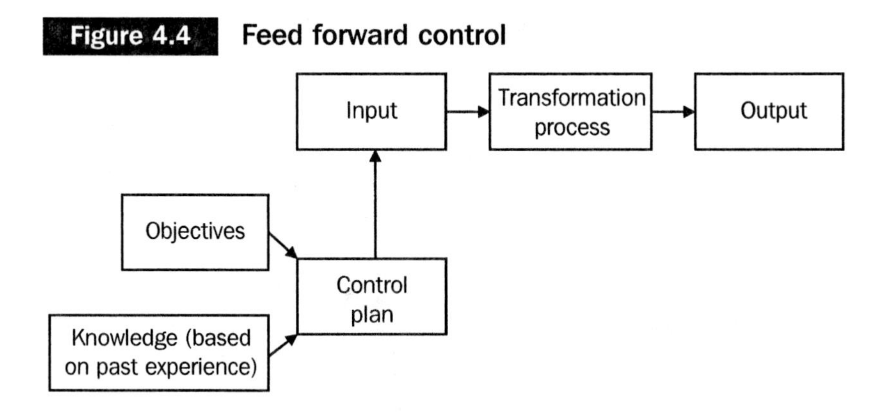

Feedback systems

Feedback control systems are therefore designed to respond to changes in the environment within which the input–transformation–output process is working. The important difference between feedback and feed forward systems is that the controller of the system determines quality levels rather than specific outputs from the transformation process. If these outputs are not to the requisite quality then action is taken to change the process so that they are. If they are already to the set quality, then no action is required. This alteration of the process and the input is known as actuation. Actuation may be an iterative process, and should last as long as required to achieve the desired quality. Feedback systems can cope with situations not foreseen by the system manager.

The key component of a feedback system is the control loop (Figure 4.5). This loop feeds information from the process that is monitoring the output back to the point at which effective actuation can occur. The loop is regulated by a comparator – the target or control output. This determines whether action needs to be taken. The feedback process takes time – known as the response time. If the response time is too short or too long, then the required improvements in the transformation process will not be undertaken effectively and the output will still fall short of the required standard. The system will become unstable: the more deviations from the normal state, the greater the likely degree of instability.

Figure 4.5 **Feedback control**

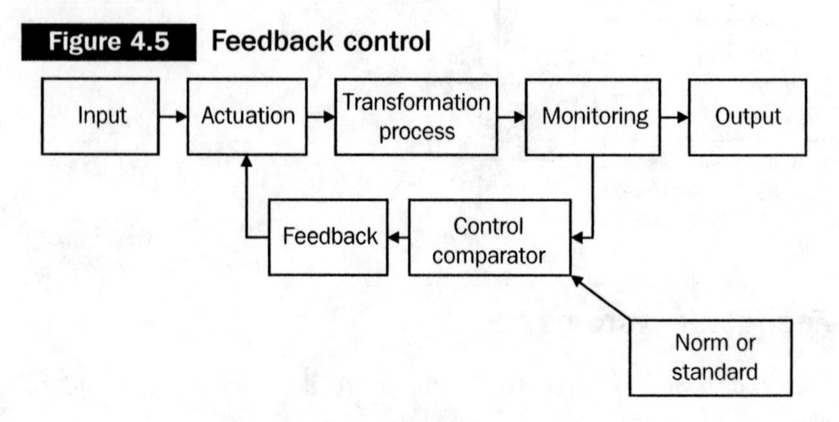

What is a difficulty and what is a mess?

In systems thinking it is usual to refer to problems (or opportunities) as being either a difficulty or a mess. A difficulty is relatively well defined. It is normally clear what the problem is and, when it is identified, what the correct solution is likely to be. It is possible to define (and measure) what success and failure will normally be over what time period and which people are involved. A mess, on the other hand, is poorly defined, with ill-defined problems and unclear solutions. Success and failure are difficult to define and differentiate. The desired objectives are not agreed or even known and no timescales have been identified. There is just a major problem that can no longer be tolerated. In reality, problems fall somewhere on a continuum between the two extremes. The better defined the difficulty, the more likely reductionist thinking will succeed. The more extreme the mess, the more some form of systems thinking will be required.

What are the boundaries of a system?

A library system will be run through the expertise of a number of staff. They will all be knowledgeable about a part of the whole. The strategy managers in particular will therefore need to be clear about 'the big picture' before they make changes in particular parts of the system that may make eminent sense to the experts in that area but which may affect the service elsewhere in such a way as to make what is offered to the users less attractive than before. In other words, they will need to know what the overall boundaries of the system are, and where the system stops and the environment in which it operates starts.

At some point, a limit has to be set to the number of component parts that can realistically be included in a system when considering possible changes. A series of boundaries has to be set. How this is done will depend on the perspective of the person or group undertaking the work. End-users, for example, are likely to have a different perspective from library managers and therefore set different boundaries from each other if asked to define a library system or subsystem. There will therefore have to be several iterations of the boundary-drawing process, taking due account of all the various viewpoints, before a clear definition of the system can be given. Say, for example, that a stand-alone inter-library loan management system is to be replaced by one that is integrated with the rest of a library's automated library systems. Although there may be a gain in efficiency as a result of the integration process, users who have enjoyed the personalised service of a specialist inter-lending team may feel that they are receiving less good treatment under the new system. They have a different perspective on the system being reviewed from those who are determining the strategy and implementing the technology.

The key objective in drawing boundaries is to determine what is part of the system to be analysed and possibly changed and what is a part of its surrounding environment. In this context, the environment outside the system boundaries is not a natural environment but only one that contains those components that are important to the system – either that affect it or that are affected by it. In order to keep the implementation process as simple as possible, the number of components within the system and the surrounding environment, together with the interactions between them, should be kept as simple as possible. The higher the level, the broader and more simplistic the system description will be. In order to carry out the task properly, the components within a given system and its surrounding environment must all be at the same level – including components that are at different levels. Including a particular element of software in a system that otherwise concentrated on overall processes, including people and hardware, rather than the whole of the software package itself is an example of mixing different levels in the same piece of systems thinking.

What are the main approaches for dealing with complex problems?

In order to solve a problem or maximise an opportunity, it is necessary to understand the problem or the opportunity fully. Without an effective understanding of this kind, strategy formulation and implementation are unlikely to succeed. There is no single 'correct' way of approaching a technology management problem. In practice, a judicious blend of techniques is most likely to yield the best results. The most appropriate approach may vary depending upon the stage that a resolution of a particular problem has reached. The challenge for the technology manager is the effective application of the available tools.

There are two basic types of approach: the hard systems approach (HSA) and the soft systems methodology (SSM). For specific technology management problems at the machine or detailed process level, a more scientific methodology might usefully be employed. However, as this book is about higher, more strategic-level problems, HSA and SSM are the recommended approaches.

What is a hard systems approach?

The HSA takes both quantitative and qualitative issues into account. The approach involves a number of activities. At all stages it is important to allow for iteration of a sequence or sequences in order to incorporate new information, not least because the completion of a later stage requires it. Although the HSA looks like a logical, well-ordered approach on paper, it becomes more complicated in reality when iterations take place, as they often do.

The approach requires two separate roles to be identified: consultant and client. The consultant need not be someone from outside the organisation, but it is important to ensure that such a role is fulfilled. At the start of the process it is important to ensure that the client's level of awareness of the problem and their level of commitment to solving it is known and understood. The problem may well be poorly defined at this stage. It may even be nothing more than a vague sense of unease on the part of the client. The HSA encourages both the definition and the redefinition of the problem until it is sufficiently focused for the later stages of the process. It also requires the consultant and the client to understand why the problem needs to be solved and why they are involved. If there is general agreement that the problem needs to be solved or the opportunity taken, then the approach can be considered as a way of handling the situation.

The first stage is to identify the opportunity or problem, to describe it and the existing system, the environment in which old and new systems operate/will operate and the systems' behaviour in given circumstances. The first stage of any project is the most crucial, and mistakes made here will affect the success of the later stages and the project overall. The project participants must therefore take care to look at all aspects of the project and the new system that will solve the problem or maximise the opportunity. As noted earlier, this requires a holistic approach that takes account of all possible perspectives. One way of doing this is to have the project participants construct a series of multiple cause diagrams. This may well reveal differences of perspective between the various participants in the project. These will have to be reconciled as part of an iterative redefinition process.

Stage two can begin once there is a clear definition of the problem or opportunity. At this point the key activity is to identify the objectives and the blockages to achieving them. In other words, those involved in the problem have to answer the question: 'Where would we like to be and what is going to stop us getting there?' This articulation process is useful for several reasons:

- It makes those involved clarify their thinking about what they really hope to achieve.
- It should bring out into the open any disagreements between the participants.
- It builds ownership of the project and the solution.
- It provides a firm basis on which to build the rest of the project, provided the process has been undertaken in a meaningful and effective way.
- Objectives and constraints can be either quantitative or qualitative. The first group can form the basis of modelling exercises; the second group will assist in the determination of the overall boundaries within which the new system will have to operate.

The objectives that are identified need to be set into the broader context of the overall mission and goals of the organisation, the programme of projects or the larger system of which the project or system under consideration is a part. This hierarchy of mission–goals–objectives will be repeated at different levels, and it is important to ensure that the appropriate level is chosen for the project under consideration.

Stage three involves the identification of the preferred route or routes by which the objectives can best be achieved. This is the most creative part of the process, and a wide range of techniques is available, as discussed elsewhere in this book.

Stage four is about evaluation of performance. The HSA stresses the need to have measures in place by which the effectiveness of the proposed way forward can be tested. Typically, measurements of effectiveness are linked to the objectives. But it is not just about knowing 'when you have arrived', but how well you have performed in getting there and whether your arrival is well timed and in the appropriate place. Several measurements are likely to be required in order to measure effectiveness and efficiency to the full. Cost is an obvious one; timeliness is another. Other measures are likely to centre upon quantifiable improvements, such as turn-round time or increased productivity in a service or technical process. The achievement of this target is both a clear and a measurable goal. These latter targets need to be set realistically in the context of the overall environment in which the system will have to operate. A target of 5 per cent, for example, could be high or low depending upon the context. Evaluation of measures used in previous programmes and projects could be useful. Were they successful? If not, why not? There are always lessons to be learned from previous activity.

There will often be a tension between the different objectives: cost and quality are the obvious ones, but particularly cost and other objectives are likely to cause tensions within a project. Given the finite nature of resources available in most circumstances, the emphasis will normally be on the reduction of cost wherever possible. Wanting to get a project finished as soon as possible and to the highest quality will drive up the costs of completion. There is therefore a need to determine the best route between low cost and high performance in the achievement of the other objectives. But even an optimal route may not be possible. A decision will need to be taken as to when and where 'the best is the enemy of the good' or where a satisfactory route is preferable to the optimal one.

Stage five looks at the possible routes to be taken in order to ensure that the project reaches a successful conclusion, as measured by achievement of the objectives and the associated target measures. The likely outcomes from each of the proposed routes can be modelled in order that a decision can more easily be taken as to which route is preferable. This modelling process is likely to concentrate on quantitative rather than qualitative inputs and outputs.

Stage six is concerned with evaluation. Once a model has been constructed, the various outcomes that will result from the alternative routes need to be assessed in terms of their effectiveness in reaching the desired objectives. It is especially important to ensure that the model does in fact reflect the real world situation in which it is to be applied. At the end of the evaluation process a number of conclusions can be reached, but there should be a single obvious conclusion that provides the most effective solution to the problem or opportunity being faced.

Stage seven is about validating the outcomes of the evaluation process. Whereas one route may be the best in terms of 'hard', quantifiable data, 'soft', qualitative information also needs to be taken into account. It is rare that one single solution will in fact perform best in all circumstances and relating to all objectives, and at this point the perspectives of all the stakeholders need to be taken into account.

Stage eight is the final implementation stage. If the HSA has been fully and effectively undertaken, then implementation should be relatively straightforward, although it will still require the development of a programme of work or a project plan.

What is soft systems methodology?

SSM has now been used in a wide range of technology management situations, including library and information environments. It is a holistic approach that includes the 'people' dimensions of technology applications, including the broader environment in which the applications will function. Almost by definition, the opportunities and problems that are particularly susceptible to the SSM treatment are likely to be those that are at the mess rather than the difficulty end of the systems thinking spectrum. This is simply because they involve people and all the different possible perspectives that they bring to the environment in which the system is likely to be operating. As with the HSA, there is a need to identify consultant and client roles.

SSM has seven distinct stages. Stage one is about looking at the problem situation. All the factors that might contribute to the problem need to be identified. At this stage the factors are simply identified in an unstructured way. They may or may not have a causal relationship with the problem. However, it is important to determine the role that you are taking in the SSM – what are your objectives and why are you involved in the problem analysis? Once you have started the process you are a participant and not an outside observer. The roles of all the other stakeholders also need to be considered at the same time. As noted earlier, they will all have potentially different perspectives on the problem.

Stage two sees the start of the analysis. Once sufficient information about the problem has been gathered, a 'rich picture' is drawn. A rich picture is a graphic representation of all the elements of the problem, as identified through the initial process of analysis. The picture can include hard and soft information. Hard information relates to facts, data, charts, etc.; soft information concerns things such as attitudes, fears, concerns and relationships. There is likely to be a feedback loop between the identification of the factors and the drawing of the rich picture as drawing the picture may lead to further identification of important factors. Ideally, the picture is drawn by all those who have a stake in the problem situation. By doing this not only should their views be taken into account but communication between the stakeholders will also be enhanced. Once the picture has been drawn, the key themes should be extracted. The themes are then briefly described.

Stage three is about determining the ideal situation and bridging the gap between the real-world situation described in the rich picture and that ideal. Systems that meet the requirements of the themes are identified. Not all the themes and their matching systems are necessarily taken forward to the next stage – only those themes that are to be a part of the conceptual model that will form the basis of the solution to the problem. It may be that all themes need to be considered, but it is often the case that one theme stands out in the rich picture as needing attention first.

Having chosen the priority theme and its attendant system, a root definition needs to be constructed. This definition should sum up the system in one sentence. The root definition aims to identify:

- Who are the victims and the beneficiaries of the system?
- Who undertakes the activities?
- What is the transformation process?
- What 'world-view' is being used?
- Who owns the system?
- What environmental constraints need to be taken into account?

Stage four relates to the building of a conceptual model of the system. The model must show how the system relating to the root definition should work – what it is and what it does. The model concentrates on activities and their logical interaction with each other. It needs to take account of the six areas listed above that are incorporated into the root definition. Having identified the activities, they need to be linked, preferably through a diagram, to form the conceptual model.

Stage five compares the models and the reality of the rich picture. The results of this gap analysis will point to where the existing system needs to be changed and where it can remain the same.

In stage six a set of changes are discussed and agreed between the stakeholders. These changes are designed to improve the situation.

Stage seven is about taking action. Here the changes agreed are implemented. Other forms of analysis may be required and a project or programme plan and management structure will be required. Monitoring and evaluation of the plan, the changes and their effects will be important. A brief worked example of the techniques applied to LIS work follows.

A worked example

System outline

This study concerns a document delivery system (Docdel) operating in a traditional academic library (see Figure 4.6). As noted in Chapter 1 and further discussed in Case study 1, Docdel systems transform reader requests into supplied documents. The primary input of these systems is the requested document and the primary output is the satisfied request. The ILDRMS (Inter-Lending and Document Request Management System) represents the transformation technology by which the request is satisfied.

Performance maintenance

Needs. Document delivery is an increasingly vital element in library provision. Reduced resources are decreasing the ability to acquire material. Regardless, fewer documents can be held on site. Yet more is being produced. Increased competition between institutions is driving up research and teaching quality, while greater student numbers (needed to

Figure 4.6 **A Docdel system**

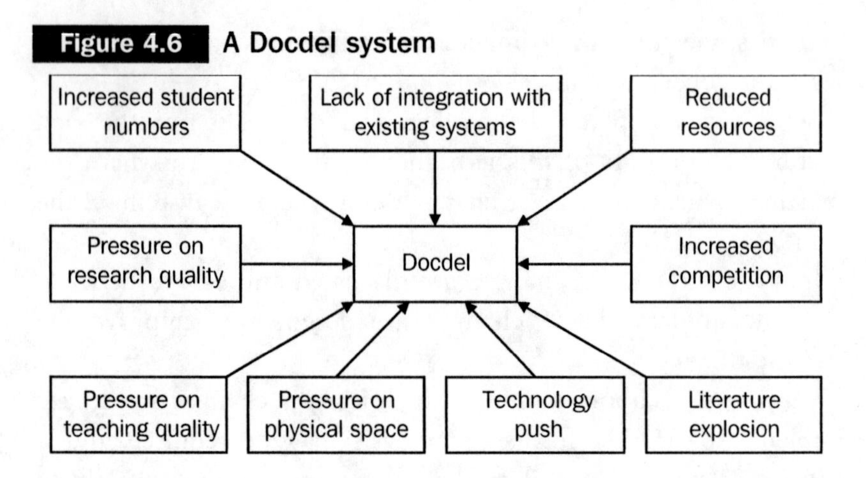

maintain the income base and to satisfy government) are reducing service quality. There is therefore a continuing need to maintain and improve system performance. This must be in the context of overall organisational objectives (library *and* university). The library has to reduce costs and increase efficiency; the university has to increase performance by the standard measures of research and teaching performance. Because a Docdel system is now so fundamental to the core business of both library and university, it is essential that any changes are in harmony with the key requirements of the users (teachers, researchers and students) and that they are successful. Library services cannot afford to fail in supporting the organisation's core business.

Successes. It is assumed that the Docdel system is subject to a service-level agreement that specifies the lowest acceptable performance in terms of turn-round time from request to receipt of documents by the end-user. In such an environment, it is expected that corrective action is taken when performance falls below the lower performance specification. The management information generated by the system informs collection development for both print-on-paper and electronic resource acquisition (see Figure 4.7). This should ensure that the deployment of resources to acquire material on-site is well targeted.

Figure 4.7 **Docdel system analysis**

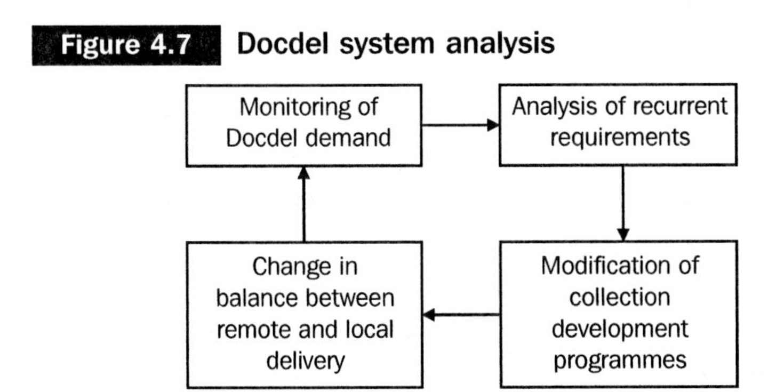

Problems. There is typically no specification of upper performance in Docdel systems (this is assumed to be an unachievable perfection) and no identification of quality terms of either primary or secondary inputs and outputs. In this worked example it is assumed that the Docdel technology is obsolescent and not integrated with other technologies or processes within either the library or more generally across the campus (e.g. with searching of bibliographic databases).

Operational improvement

Needs. There is a need to develop proper benchmarking against best practice in the field. This must include not just delivery times but also quality measures such as actual contribution of the supplied material to teaching and research quality and student satisfaction. There must also be a review of the cost structure of both the Docdel service and the library service overall. Could a re-engineered Docdel service reduce overheads for the library and the university as a whole? For example, a more effective Docdel service could reduce the need for an expensive, on-site library estate. Given the fast pace of technological change and the pervasiveness of the Web, there is a need to develop new Docdel products that capitalise on the Web's popularity and potential, as well as the widespread connectivity to, and acceptance of, the Internet by end-users of all kinds.

Successes. The Case Study library has begun to develop a Web-based integrated search–retrieve–delivery–manage and monitor service that is fully electronic. The service is a collaborative venture with other libraries, document supply organisations and a software supplier. The partnership has been very effective in terms of prototyping and identifying major needs for the development of a full product.

Problems. It is not thought that there are any obvious benchmarks by which new forms of integrated Docdel can be evaluated. It is also clear that quality is more difficult to achieve with the increased availability of so many sources of both bibliographical information and actual documents. The proposed new integrated system is an innovatory rather than an incremental change. It requires major re-engineering of existing services and processes and a culture change on the part of both library staff and users. There may be an unwillingness to embrace change; some staff may even fear for their jobs. The new system is taking time to introduce 'for real'. In the meantime, alternative systems and services are emerging that allow end-users to use the Web to bypass the present library provision. There is the possible threat of redundancy of the 'old-style' central Docdel service, however technologically sophisticated or high in quality.

Proposed system changes

Preferred methodology

The SSM has been selected in order to identify beneficial changes. This is because:

- Strategic decisions need to be taken (what is the future of library provision? where does electronic Docdel fit in?, etc.),
- Several objectives need to be met.
- There is a high degree of uncertainty (what will future provision look like?).
- Not all opportunities are known and understood.
- The problem is not well structured or defined.

- Success and failure will be difficult to define.
- Not all the components of the system can be controlled (e.g. staff and user behaviour, e-document publishing trends).
- There is a large amount of human involvement (effectively the whole of the organisation in some form).
- There are many different perspectives (e.g. staff, users, suppliers).

A systems approach would break down the overall problem into its constituent parts. However, in this case, one cannot assume that solving these individual problems will mean that the whole problem will be solved. While an HSA looks at both the quantitative and the qualitative issues, it would be more appropriate if the problem were being defined as relating to the Docdel system only. The approach would then be very useful because it could help to change the system for the better. However, because there is so much uncertainty, the view is taken that there is a 'mess' rather than a 'difficulty' to be sorted out. An SSM approach is therefore recommended. If the system-specific objectives and difficulties become clearer, then there could be a return to an HSA. For the time being, there is a need to look holistically at the whole question of Docdel in its broader context.

Essentials of working

Given this last factor, the SSM analysis must comprise the collection of both hard and soft data. Ideally, all the information will be analysed and fed into a 'rich picture'. There are two particular advantages of the SSM approach at this stage: firstly, it offers opportunities to gather information about perceptions, relationships and other 'people' aspects; secondly, it allows the people themselves to be involved in the analysis. This is likely to be useful at a later stage when solutions are being implemented, especially where people are the perceived blockage to change. In addition, the role of the 'investigator' is clarified at an early point in the process.

In the case of the Docdel problem, a rich picture might include the following elements:

Hard data

- Management information (delivery times, transaction loads, satisfaction rates, document costs)
- Publishing trends
- Technology trends
- Supplier data (documents, technology)
- Comparator/benchmark information (where available)

Soft data

- Perceptions of different groups within the organisation (senior management, users, library staff, etc.)
- Expert opinions (e.g. through Delphi exercises or, more simply, desktop research)
- The investigator's own perceptions (especially regarding issues, concerns, etc.)

The rich picture should throw up key themes, such as:

- increasing moves towards e-publishing only;
- reductions in the unit of resource available;
- devolution to individual units as cost centres/strategic business units;
- increased (total?) end-user control of access, budgets?
- library as intermediary, account manager rather than provider;
- demise of print-on-paper provision (at least for journals);
- library staff fearful of redundancy/deskilling.

It should also allow the identification of relevant systems, including ones to:

- allow library staff to embrace new technology and e-publishing;

- ensure that user needs are identified, understood and satisfied (within the available resource base);
- facilitate the integration of traditional and technology-based systems to provide seamless end-user access.

This can then be integrated to form a single system (see Figure 4.8), given that the three (sub)systems above are interrelated. From this the key aspects of the system can be identified (see Table 4.4).

Figure 4.8 **Docdel system interactions**

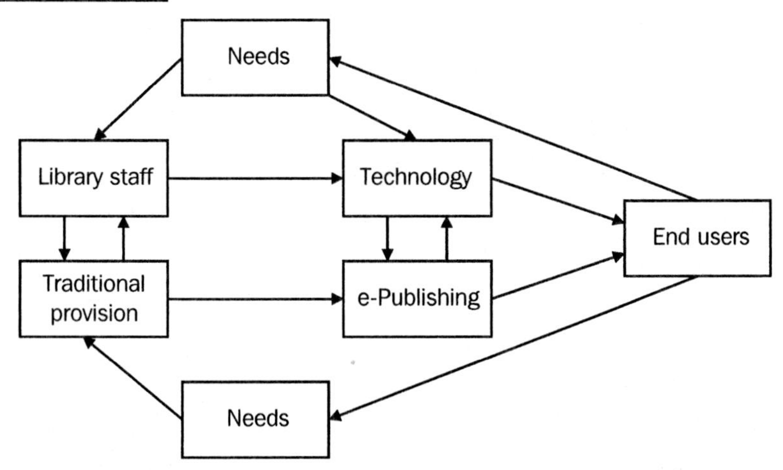

Table 4.4 **Key aspects of a single system**

Customers	The academic community (staff, students, administrators)
Actors	Library staff, senior management
Transformation	Changing both processes and services to accommodate and exploit new technology and e-publishing to the institution's best advantage
World-view	The library service needs to change in order to ensure that it remains central to the organisation's operations and that it remains competitive
Owners	Senior staff within both the library and the institution
Environmental constraints	Need to ensure that end-users receive the right document at the right time within agreed resource parameters

The 'root definition' can then be constructed, as for example:

> *A system to ensure the integration of new technology within the library, and with the positive support of staff to enable the institution to exploit e-publishing to the full, providing easy and seamless end-user access to targeted resources, as identified by senior library staff and university management.*

This is followed by a conceptual model (see Figure 4.9) based on front-line activities such as the following:

- Identify problems and opportunities (managers/library staff).
- Plan changes to best effect (managers/library staff).
- Innovate (managers).
- Involve staff and users (managers/library staff).
- Ensure compliance with both user needs and set standards (library staff).
- Monitor technology and sector key changes (managers).

Figure 4.9 Conceptual model of front-line activities

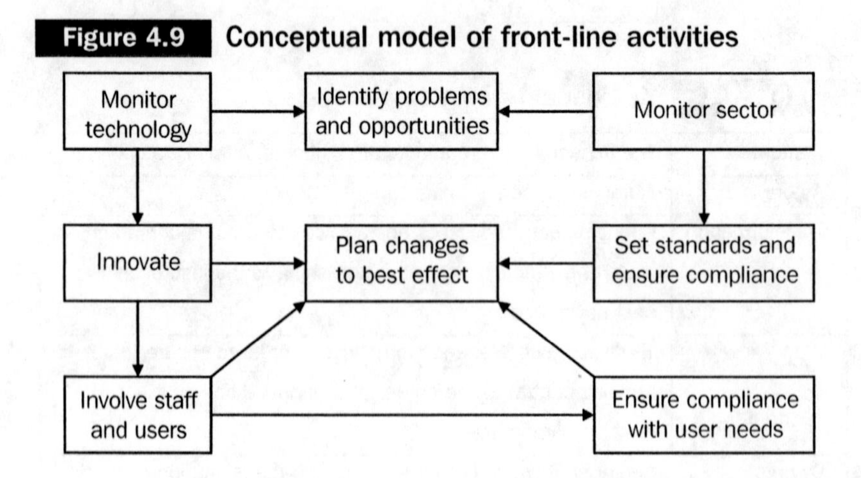

This model will require iteration and refinement until it can be related back to the rich picture that began the exercise. For example, at the moment, it stops at the planning stage and there will be a need to consider both detailed actions and their implementation/evaluation. In addition, the elements of the model are very general and need development. Monitoring the sector, for example, could be broken down into the following:

- Identify likely changes in national document delivery models.
- Forecast technological innovation among major academic publishers.
- Review developments in other universities and other countries.
- Consider collaborative arrangements.
- Identify possible sources of pump-priming funds.

Discussion between key players (library staff, users, key suppliers, IT staff, senior managers) should then be facilitated in order to determine what changes need to be made in order to benefit the organisation. This process should also enable real participation, involvement and ownership to succeed. The main changes might include:

- procurement of a new integrated Docdel system;
- collaborative arrangements with other (regional) universities regarding joint electronic access;
- staff training and awareness sessions;
- re-engineering of existing processes;
- new duties and responsibilities, etc.

Likely outcomes

It is difficult to be precise about likely outcomes at this stage, given that SSM is an interactive, iterative process. If the process is to be successful, however, outcomes should include:

- a greater degree of certainty about the best future strategic direction;
- a clear definition of the problem (a 'mess' has been turned into a 'difficulty' or a series of 'difficulties' that can be handled by using other methodologies – e.g. an HSA);

- a set of objectives that are capable of implementation;
- a broad degree of ownership of the problems, the objectives and the solutions (because there has been iterative and facilitative involvement of all key players);
- new ways of working (in terms of both new processes and also a willingness to use SSM in the future);
- new opportunities for developing services to best effect.

Now that the SSM has revealed the key difficulties, it is necessary to tackle each of these. What began as a technical problem seems to be evolving into largely a 'people' problem. This is often the case, although the key technological drivers must also remain at the forefront of managers' minds. However, a key system required is one that allows library staff to embrace new technology and e-publishing. Library staff will have to own the changes and the new systems. If the staff are such an important element in the success of the changes, then what are the key issues that need to be identified? What will make them resistant to change or positive about a new Docdel system, and especially one that threatens to deskill them or make them redundant?

A multiple-cause diagram has the power to determine cause and effect (see Figure 4.10). Creating such a diagram forces the manager to think through the main elements of the problem and to ensure that, as the person developing the diagram, he/she is fully aware of what the problem really is.

In the example of library staff, given above, it could be argued that the resistance to change is simply because of their conservative attitudes. Many of the older personnel were trained at a time when library automation had impacted on 'housekeeping' activities and nothing more. However, the diagram reveals a much more complex picture. Although it is true that the traditional training that staff have received means that there is a degree of ignorance about the new technology, this leads to underlying fears of uncertainty, deskilling and job losses. The speed of technological change makes it difficult to keep up anyway, and the fears

Figure 4.10 **Multiple-cause diagram of the Docdel system**

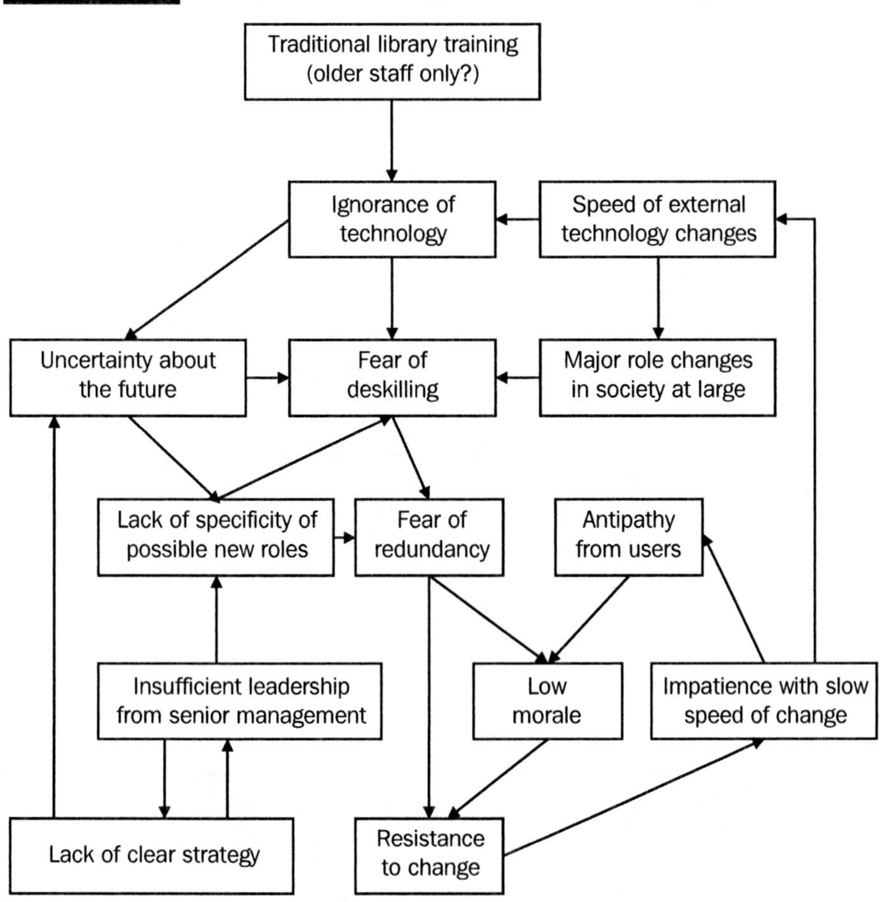

are reinforced by the changes that the staff sees in society. 'Will it happen to me?' they say.

These changes are also affecting the way users see the services. Web technology allows them to be much more in control and library staff are seen as blockages rather than helpful intermediaries as they once were. The resulting antipathy of users, combined with the fear of redundancy, leads to low morale and hence a resistance to change ('we need to hold on to what we've got').

Some of the external factors are not under the control of management, though an effective response has to be made. But Figure 4.10 reveals an area where management could and should be proactive. The diagram shows that in this case, uncertainty about the future is not just as a result of technology ignorance or the speed of external change. It is also as a result of the lack of a clear strategy within the institution and, in consequence, the lack of leadership from senior management (though the lack of leadership also impacts strategy in reverse). This in turn leads to the lack of clarity about possible new roles and hence reinforces the fear of staff regarding deskilling and redundancy ('if I don't have my old job, and there's no new job being created, then I'm out').

Without the diagram, it would have been easy to criticise staff for their unwillingness to change. It is now clear that they have good reason to be resistant, but the management has a number of opportunities to turn the present negative situation into a positive one.

Summary

This chapter has looked at some of the more specific techniques for dealing with strategic problems. These are all based on systems thinking and the idea of the model and modelling. The aim is to link operational and strategic management in realistic and effective ways. The concepts of reductionist and systems thinking were considered. A systems thinking approach adopts the broader view that strategic management requires, although reductionist thinking has its uses in more specific problem-solving situations.

First of all, the problem has to be defined. Several techniques have been recommended and described. Boundary examination does just that, seeking to create an area in which the problem can be contained but not constricted. The 5Ws and H technique is a simple question-and-answer approach to problem definition that can lead into the more sophisticated dimension analysis.

It was suggested that underpinning 'real-world' problems were generic concepts, whose identification would provide not only clarification of the specific problem to be solved, but also add value to the more general strategic planning process.

The rest of the chapter looked at systems thinking in more detail, beginning with feedback systems, types of problem, and the two main methodologies for dealing with complex problems: hard and soft systems approaches/methodologies.

Bibliography

Anstey, P. (2000) *C & IT Skills: Developing Staff C & IT Capability in Higher Education*. Norwich: University of East Anglia.

Heeks, R., Mundy, D. and Salazar, A. (1999) *Why Health Care Information Systems Succeed or Fail*. Manchester: University of Manchester Institute for Development Policy and Management.

Van Gundy, A.B. (1988) *Techniques of Structured Problem Solving*. New York: Van Nostrand Reinhold.

5

Strategy implementation

Programme and project management

Programmes and projects are the ways in which strategy is turned into reality; they comprise the main way of actually 'making things happen'. We are looking here, then, at the bridge between the high-level strategy and the operational or the day-to-day. However, although programme and project management relates largely to practical matters, it is also about making changes to existing products, services, structures or ways of working. This requires programme and project managers to interact with other parts of an organisation and for senior managers – typically the people who sponsor programmes and projects – to ensure that the work that they have commissioned is well integrated with the organisation both operationally and strategically.

Effective strategy formulation must ensure that when the strategy is implemented, the three key objectives of the classic project management triangle are met in order to produce a result:

- on time
- to budget
- of the required quality.

These three criteria are the fundamental building blocks of the techniques discussed here. They underpin all implementation programmes and activities. The challenge is to balance the conflicts between the three variables. A project that is on time and to budget, for example, may not be of the required quality; only the injection of resources additional to the original budget or the extension of the project beyond the original deadline may actually produce the required quality once work has actually started and the true requirements have been assessed.

What is a project?

A project can be defined as 'a group of connected activities with a defined starting point, a defined finish and need for a central intelligence to direct it' (Taylor and Watling, 1973). The 'central intelligence' is the 'glue' that binds the various elements of the project together. This is typically supplied by a project manager, backed up by a project management structure appropriate to the requirements of the project. The project manager's role is a challenging one. It requires the post-holder to plan, schedule, direct and control a wide range of resources not only within constraints that are set by the project sponsors (the three basic criteria noted above) but also in the context of external environments that present an additional, and often uncertain, set of challenges.

What is a programme?

A programme is typically a cluster of projects that together are intended to meet the aims of the commissioning organisation. The projects within the programme will need to be integrated with each other and with the overarching needs of the parent body that is controlling and resourcing them. Programmes, like individual projects, must be led and managed. This is normally done by a programme manager, backed up by some form of programme management board.

Why are programmes and projects undertaken?

A project, or a programme of projects, is typically undertaken in order to carry out the aims and objectives of a given strategy. A change or transformation will have occurred at the end of a project or a programme. This could be the introduction of a new product or service, or the implementation of a next-generation technology or some other improvement that is designed to improve efficiency or effectiveness within the parent organisation and to make it more competitive. It is the finite nature of programmes and projects, together with their fundamental aim of bringing about change, that makes them different from the normal ways of working – ways that are designed to cope with routine and predictable situations. Because the environment in which any institution operates is constantly changing – and not least through the continual upward spiral of technology advancement – the benchmarks by which projects and programmes will be judged will vary over time and in the context of the generic or sectoral environment in which they are being carried out (Clarke, 1999). It is thus crucial that programmes and projects are 'fit for purpose' with the overarching strategies that they are intended to support. This requires the development of intermediate-level aims and objectives that on the one hand link with the higher strategic levels and on the other hand give those who are implementing and managing the programmes/projects sufficiently concrete aims and objectives with which to work and on which they and others can judge the effectiveness of the projects being carried out.

How are programmes and projects best organised?

There is no single way of organising a programme or a project to best advantage. A number of factors will influence the ways in which a project or programme is managed. Context, culture, external environment and organisational priorities are some of these. In the context of this book, of course, technology is a key driver in project and programme management. Some programmes and projects will be directly concerned with the introduction of a new technology; others will use a technology to facilitate the management of the programme or project.

There is one element of programme and project management that tends to be common to most activities of this kind, regardless of the particular context. It is highly likely that the management will be organised as some kind of matrix, with the programme or project manager managing, or at least interacting with, a wide range of staff from different areas of, and levels within, the organisation.

In order for programmes and projects to work successfully, it is essential not only that there is a highly competent manager in charge of all aspects of this matrix, but also that this person has the authority to act and has the support of the senior management within the organisation, especially where the desired changes that form the output of the programme or project are controversial and not universally welcomed by all the stakeholders. The role of the programme or project manager in achieving success is discussed further later in this chapter. Suffice it to say that, because of the uncertainties and changing aims and environments that surround most programmes and projects, they

> ... must be actively managed in order to allocate resources and instil a sense of urgency. It demands a person who understands the totality of the project in relation to its business objectives,

who is dedicated to its success, and who is in control of the resources needed to achieve the end result. Whilst a good planning and control system is essential, it can only provide a framework for decision making and supply the information required for managerial intervention. The greater the degree of change from normal practice, the greater the number of variations from the plan and the greater the demand upon management.

Because there is an urgent need for remedial action it cannot be left to the deliberations of a committee. This is the rationale for a system of project management which enables one person to take timely decisions with the minimum of interference. This person's position is analogous with that of the champion, but whereas the champion assumes the role him or herself, the project manager is appointed to implement a change which has already received the support of management.

(Twiss and Goodridge, 1989)

What are the main stages of a project?

Every project will have a number of stages or phases. These will typically be delineated by key milestones or objectives. These need to be met at various critical points in the project's phasing. The chances of completing a programme or project successfully will be significantly increased if milestones are clearly specified so that there is no room for doubt or dispute as to when and whether they have been completed. The five-phase model shown in Figure 5.1 (adapted from Thamhain, 1990) sums up the key stages of a project.

Figure 5.1 **Key stages of a project**

Conception

Development of initial broad objectives, scope, constraints, needs, feasibility. May be underpinned by a stakeholder and/or environmental analysis as well as a financial assessment – likely budgets and return on investment. There will also be a need to estimate the time required to complete the project.

↓

Definition

If the output from the conception stage is approved, then the project is defined in full. The main stages of the project are identified and broken down into manageable elements or work packages, each of which has a set of deliverables, tasks and dependencies. Resource requirements and timescales are also identified. This is the stage at which technologies to be used will need to be identified. What will be the inputs, outputs and transformation processes and how are they to be changed as a result of this project?

↓

Organisation and start-up

Once approval is obtained from senior management, the project can begin. At this stage, all the necessary 'housekeeping' activities that are required before the project can begin must be undertaken. This will include issues such as technology supply and support, staff recruitment and training, organisational structure. There will be a need to determine both quality standards and performance indicators and the means by which these can be achieved.

↓

Execution

This is the phase during which the project is actually carried out, as far as possible according to the full project plan and, ideally, it ends with the delivery of the product, service, improvement or other desired change. The extent to which the project is successful during this phase will depend in large measure on the effectiveness of the management, both in day-to-day aspects of the project and in the management of the 'exceptions' that may cause the project to run behind time or below standard or over cost.

↓

Phase-out

The project is wound up at this stage. All deliverables should have been completed and handed to the project sponsors. Formal project review should take place and any lessons learned summarised. There may be follow-up projects that are identified also at this stage.

In reality, most projects will be much more complex than the basic five-stage model suggests. The diagrams reproduced later in this chapter, for example, suggest a number of iterations will be required at various stages of the project and in relation to particular aspects of the desired aims and objectives, such as the technology to be used. In addition, it will be important to ensure that the key risks to successful completion have been minimised and that there are back-up plans in place in the event of the programme or project under-achieving at any point. Risk management is discussed in more detail later in the chapter.

Project management methodologies

Most programmes and projects adopt a methodology in order to structure the way in which they are managed. There are many such methodologies available and as many books on how to apply them effectively. A popular methodology for technology-based projects is PRINCE (PRojects IN Controlled Environments). PRINCE provides a framework for project and programme management under a number of key headings:

- Organisation: teams, roles, relationships
- Plans: technical, quality, resource, time-based
- Controls: standards, monitoring, feedback
- Products: hardware, software, documentation
- Activities: management, decisions, research, development.

PRINCE offers a very sophisticated and detailed approach to programme and project management in the IT sector. For many projects, it is far too big a methodology to be effective. However, the principles of the approach can be adapted to smaller-scale projects and programmes (Lewis, 1995).

Lessons learned

Very few projects will go exactly to plan; there will always be 'room for improvement'. Methodologies such as PRINCE encourage programme and project managers and their sponsors to identify the 'lessons learned' from completed work, whether successful or unsuccessful. This chapter incorporates a number of lessons learned from some major ICT-based projects in which I was involved, and also includes the views of a number of other senior programme and project managers. The results of such reviews are likely to be of use not just in future programme and project work, but also in relation to the development of strategy, with particular relevance to organisational capacity to implement strategic aims and objectives. If an institution has a poor reputation for programme management, for example, it is highly unlikely to be able to turn its aspirations into reality without major organisational change. In such a case, the institutional strategy will have to concentrate on capacity building as much as change management and technology improvement or innovation.

Programmes and projects in context

It is important to ensure that programmes and projects fit into the overall institutional aims, objectives and management structures. Because of the matrix approach noted earlier in this chapter, it is important to ensure that finite projects fit into routine operations of the organisation while they are in progress and, once completed, that their results can be integrated into the future workings of that organisation. There are many examples (discussed in this chapter and the case studies) of projects that were successful in terms of delivering their objectives but unsuccessful in terms of the actual take up or 'embedding' of their outputs. In addition, programmes and projects have the capacity to unbalance an organisation in terms of day-to-day operations, especially where staff are seconded to the special work, with perhaps a responsibility additional to that of their colleagues. This requires careful handling if the project output is not to be prejudiced because of local enmities. More fundamentally, it is essential that programme and project work is complementary to current activities rather than detrimental to them, in terms of both resource allocation and management time devoted to what may be perceived as non-core business. Much will depend on the particular environment in which the organisation is operating and the extent to which project and programme activity is crucial to future viability.

Success: an introduction

The 'strategic trap is embarking on a project without first assessing the full consequences of its success' (Roussel et al., 1991). Any approach to the strategic or the operational management of technology needs to take account of previous success or failure in the area. However, there is no widely accepted definition of 'failure' or 'success' even within LIS work. It is not always easy to forecast, manage or determine the extent to which a given programme or project has succeeded or is likely to succeed. Where research and development is concerned, negative results are typically deemed to be as useful as positive ones. In Joint Information Systems Committee (JISC) calls for proposals for projects, for example, reference is made to the 'recognition that in groundbreaking work there may be failures as well as successes, but that all such experience can provide valuable information for the [UKHE] community' (JISC, 2002). This is a standard wording for JISC initiatives.

Much of the rest of this chapter grew out of actual experience of technology projects. The information summarised there is a distillation of a great many 'lessons learned'. Data and knowledge extracted from past practice, successes and failures form one of the most important resources available to the technology manager tasked with the implementation of strategy. What did and did not work? What is likely to work in the future, with a given technology, in a particular environment?

Stakeholder perspectives

Different stakeholders have different perspectives of success, failure and process that all need to be taken into account (Kirby, 1996; Fowler and Walsh, 1999; Gray, 2001). There is a need to be clear about what constitutes success and to understand *how* and *why* a project is successful or unsuccessful in order to reduce failure rates and maximise the considerable investment now being made. Traditionally, 'success' is the completion of a project on time, to budget and to agreed quality (Zwikael et al., 2000). It is the delivery of what was promised, expected and hoped for – a product or service that does what the project instigators said it would do and which yields benefits to those who subsequently use it and at whom it was aimed (Bradley, 1997). Assessing project success in a research and development framework is about more than just the standard 'iron triangle' of cost, time and quality (Lopes and Flavell, 1998; Atkinson, 1999). The question of success in project management is a fundamental and long-standing one that pre-dates IT-based developments. There are practical, systemic and philosophical elements. Not only are there challenges concerned with identifying what variables might cause problems in given circumstances, but there is also the added difficulty of the *interaction* between the variables (Nellore and Balachandra, 2001).

Reasons for failure

Software project management has particular difficulties that need to be addressed (McDonald, 2001; Vandersluis, 2001). Project failures can stem from attitudes, actions, technology, or all three (Jelinek and Schoonhoven, 1990; Noori, 1990). Social and cultural issues have a particularly large part to play in project success or failure (Gray, 2001).

Projects may fail because they are too radical (Sundbo, 1997) or because there is an inappropriate balance between the push of the technology and the pull of the market in which it is being introduced (Kumar et al., 2000). 'Technocratic utopianism' (Davenport, 1994) can get in the way of pragmatic business sense and projects can be mismanaged despite, or perhaps because of, the methodology used (Twiss and Goodridge, 1989). IT projects – as seems to have happened in e-Lib – tend to 'escalate' out of control (Drummond, 1999). Without standards for effective project management, failure, however defined, is likely (Johnson et al., 2001; Sampath, 2001).

Success criteria

Cooper and Kleinschmidt (1987) looked at the criteria for success with particular regard to new product development. They found that success (or failure) is governed by the nature of the interaction between the market environment and the appropriateness of the new product strategy and its implementation. This interaction takes place within the context of the technology life cycle. Cooper and Kleinschmidt concluded that the key success factors relating to the introduction of new products, in order of priority, were:

- strategic consideration of the product design (features, attributes, advantages);

- pre-determined strategies on the product launch (target market, price).

The model in Figure 5.2 aims to summarise the key aspects of the challenge of forecasting success or failure. The measure or measures by which success or failure can be assessed are an amalgam of project outcomes with definitions of success and a range of appropriate performance indicators. If one knows what success means, and has ways of measuring it, then one can determine how successful a project has been. If one can trace back to the beginning of a project and identify what made it successful or unsuccessful, then those variables that affect project outcomes can be assessed and measured. They can be included or strengthened in future projects in order to improve success rates.

Figure 5.2 **Forecasting and measuring project success**

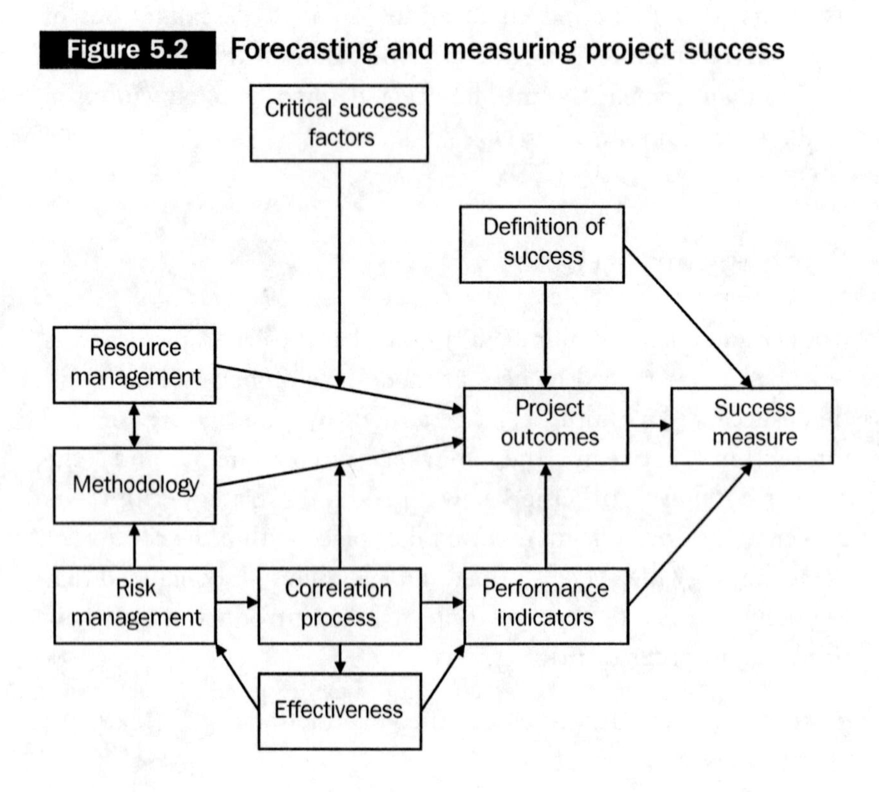

Risk management

Risk management is fundamental to the success of any project (Chapman, 1997; Jaafari, 2001). The model postulates that the effectiveness of risk management techniques and of development and usage of performance indicators is fundamental to forecasting success. Together, risk management and performance management underpin the correlation process. The more effective the risk management process and the more correlated it is with desired outcomes, the greater the chance of success. In managing risk, an organisation can then apply the four 'T's: terminate or stop the activity; treat or control the risk; transfer (e.g. insure) the risk; or take (accept) the level of risk identified.

How does one measure and forecast success in IT development projects? Are there generic definitions of success and failure that can be applied? What are the major variables that can affect success rates? How can they be correlated? Recent literature suggests that there are ways of forecasting project success (Baccarini and Archer, 2001; Davis et al., 2001; Vandersluis, 2001) and the key challenge for managers is to develop approaches appropriate for their circumstances.

Risk management should begin at the conceptual stage in order to maximise the chances of success (Uher and Toakley, 1999). But it has to be integrated with the rest of the project (Jaafari, 2001). Nor is risk management an exact science. It needs to take account of specific contexts (Ward, 1999). The literature is divided on what is most appropriate (Raz and Michael, 2001; Stewart, 2001). Risk management processes need to be expanded and enhanced (Pender, 2001), not least to include risk associated with the nature and composition of the project team (Williams, 1997).

Resource management

Management of resources available to the project is a key part of the process. There are generic skills concerned with managing resources that are essential to the success of a project (Johnson et al., 2001). These consist of political nous (Pinto, 2000), good 'timing' (Thoms and Pinto, 1999) and business process re-engineering (BPR), where it is especially important that resources are deployed as flexibly as possible (Dey, 1999).

Project manager workloads need to be optimised if resources are to be managed effectively and project outcomes are to be as desired (Kuprenas, 2000). As importantly, clients, project teams and functional managers should be involved at all stages (Jiang, 2000). This needs to be handled carefully if it is to be effective (Vadapalli and Mone, 2000). It might best be done through the use of total quality management (TQM) techniques (Orwig and Brennan, 2000).

Critical success factors

Critical success factors (CSFs) need to be taken into account. In the model discussed here, the CSFs are placed between the inputs of methodology, risk and resource management and the project outcomes, with a strong link to the correlation process described earlier. If the key CSFs for a given project have been recognised and are catered for at the beginning of the project, then there is likely to be a greater correlation between inputs and successful outputs. It has been argued, for example, that one of the critical success factors in the development of the information environment in UK higher education has been the fact that the material made available has been free at the point of use.

CSFs can be categorised on the basis of experience across a number of industries and used to improve success rates (Johnson et al., 2001; Nellore and Balachandra, 2001). The CSFs drawn from other sectors include: leadership style (Thite, 2000), team communications (Thomas, 1999), the ability to iterate and be flexible (Ambler, 2001) and the ability to spot trouble early and take decisive action, including project closure if necessary (Feldman, 2001).

Project design and management

Good project management cannot make up for a poorly designed project; conversely, a good project is unlikely to be made better by good project management (Munns and Bjeirmi, 1996). Strategic viewpoints are crucial if success is to be achieved (Artto et al., 2001). There is still a generic need across a broad range of technology management sectors for new models for capturing project requirements to be developed (Hvam and Have, 1998). There is typically a need for business-led exit strategies, especially in public sector R&D though these are often noticeable by their absence (ESYS Consulting, 2001). They are required if projects are to turn into income-generating products or services (Jaafari, 2000).

Measuring and forecasting success

The model shown in Figures 5.3–5.5 develops the concepts underpinning the 'real world' problem of measuring and forecasting success, drawing upon the relevant literature, with special reference to IT project management. Many headings are as in the first model: project management, resources, methodologies and risk management, definitions of project success, performance indicators and forecasting. The major change is the technology management context. Much of IT project management is about *innovation* management (Twiss, 1992; Twiss and Goodridge, 1989). The key link between project and innovation management and the forecasting, measurement and delivery of success is risk management.

Figure 5.3 **Project success: the conceptual subproblem**

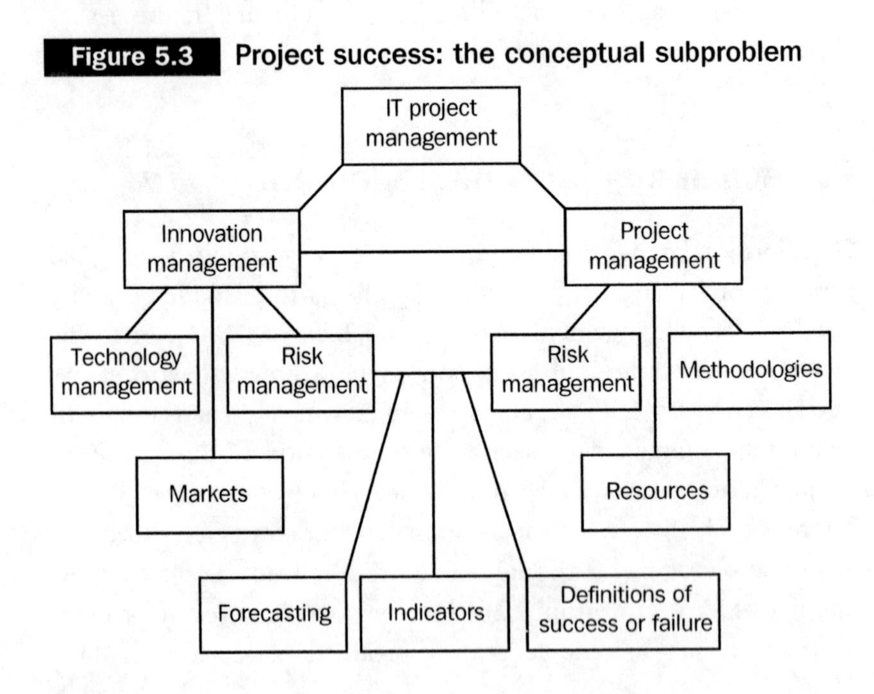

Gaps between reality and conception

The Tavistock Institute (2000) found that JISC's e-Lib projects lacked definition at project start-up – a common phenomenon (Webster, 1999). Heeks gives examples from the UK National Health Service – another public sector organisation – where 'technocratic utopianism' (Davenport, 1994) leads to disparities between requirement, expectation and actual outcomes. Project requirement documents should make the maximum effort to close this gap between reality and conception (Heeks et al., 1999).

Figure 5.4 Reality vs. conception (1)

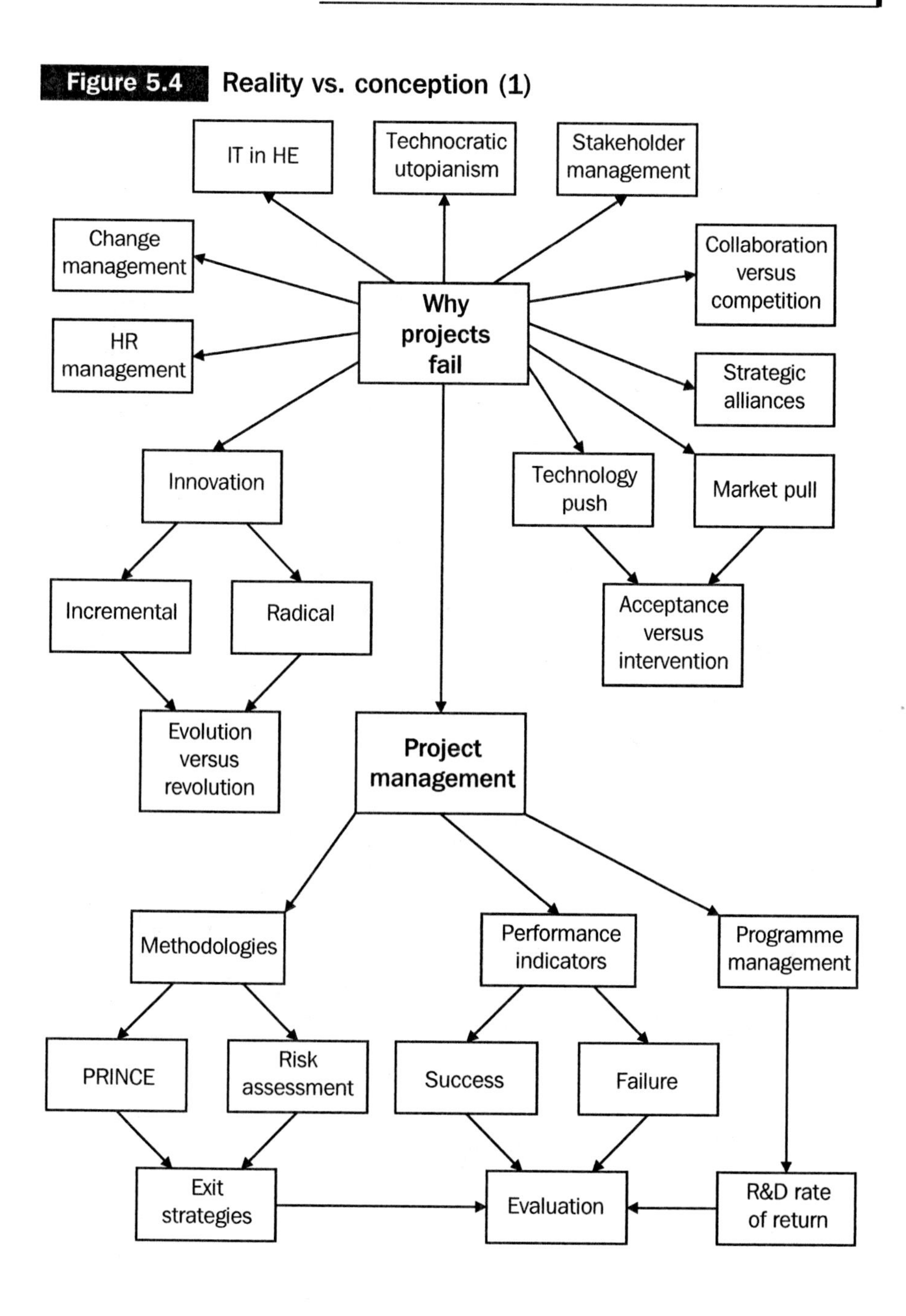

Figure 5.5 Reality vs. conception (2)

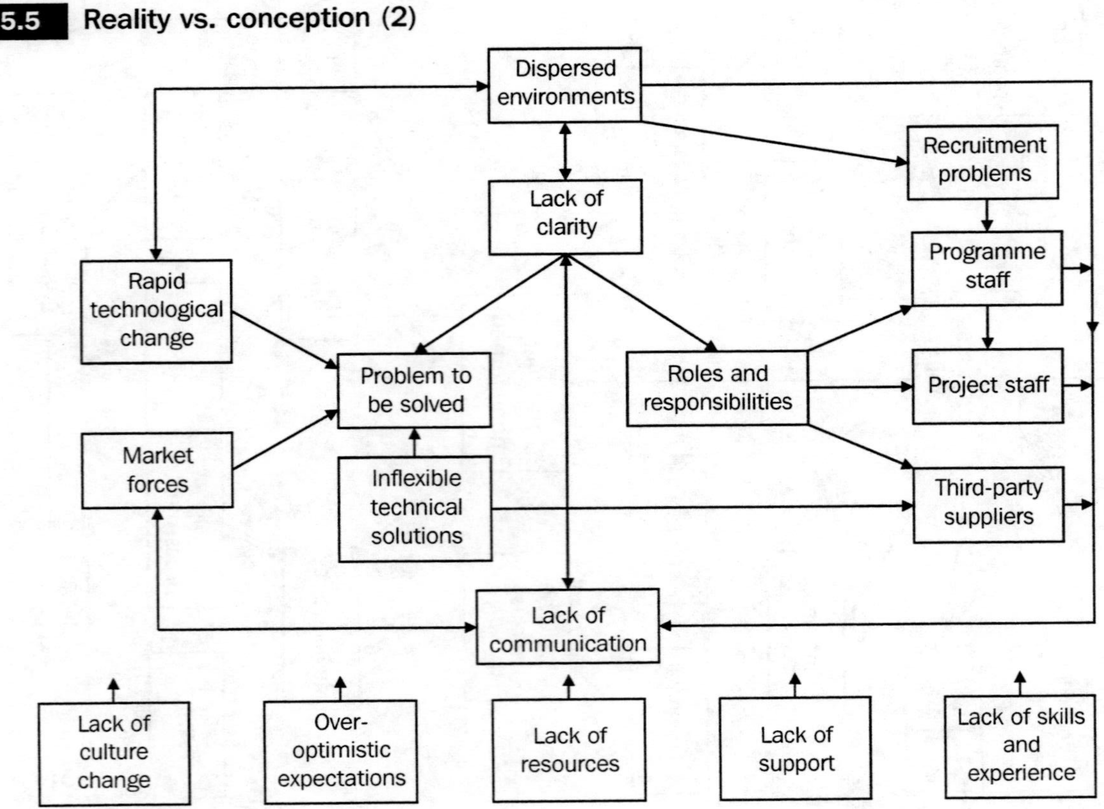

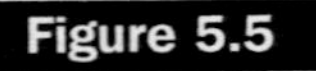

Successful projects: a model approach

The models shown in Figures 5.6–5.17 aim to summarise all the elements of a successful project. The first of these includes the key attributes of a successful project, the key drivers or critical success factors and the ways in which risk is best managed. This model is set within the context of the need for adequate resources, an appropriate project management methodology, an effective use of technology (whether as the agent of change or the output from the project) and a realistic understanding of the environment in which the project is being undertaken.

The remaining models look at each aspect of successful programme and project management and aim to provide a complete summary tool kit of how best to assess the likelihood of success within programme and project management where there is a high level of technological input and engagement.

Figure 5.6 **Key attributes of a successful project**

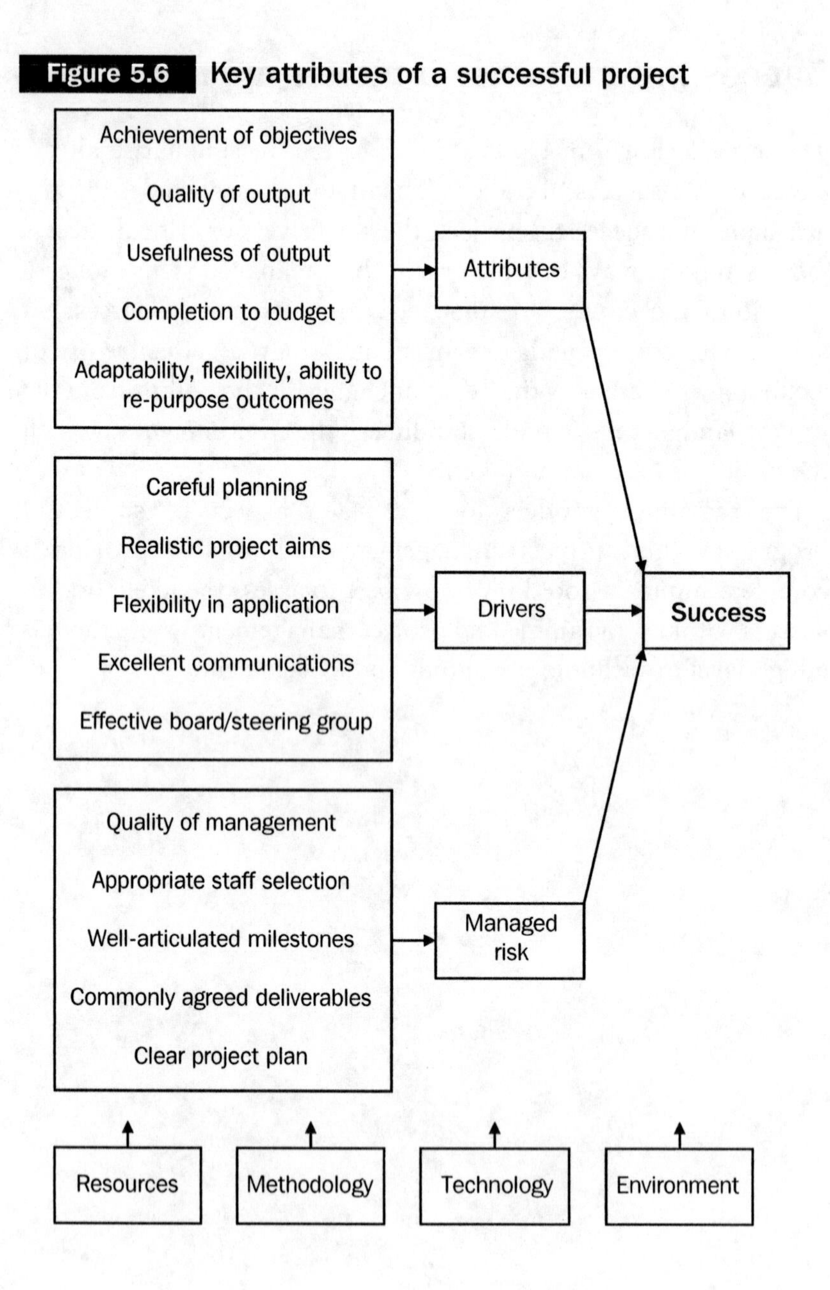

Figure 5.7 Management effectiveness

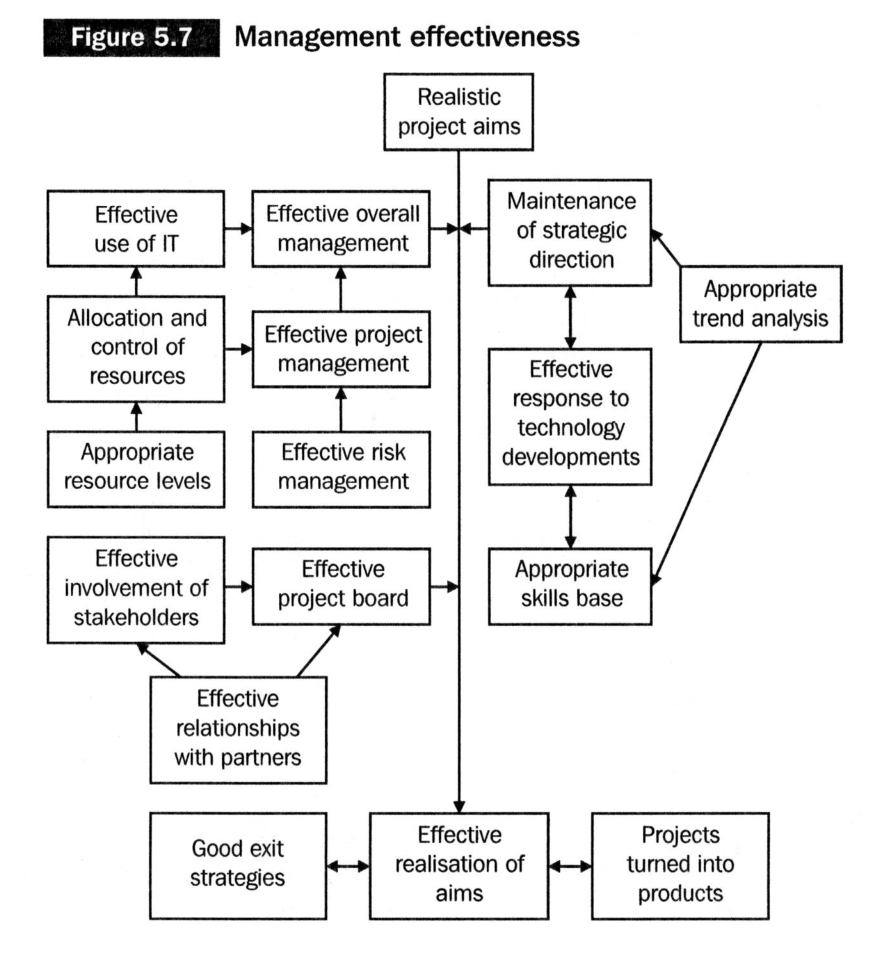

Figure 5.8 Success measurement and forecasting: key headings

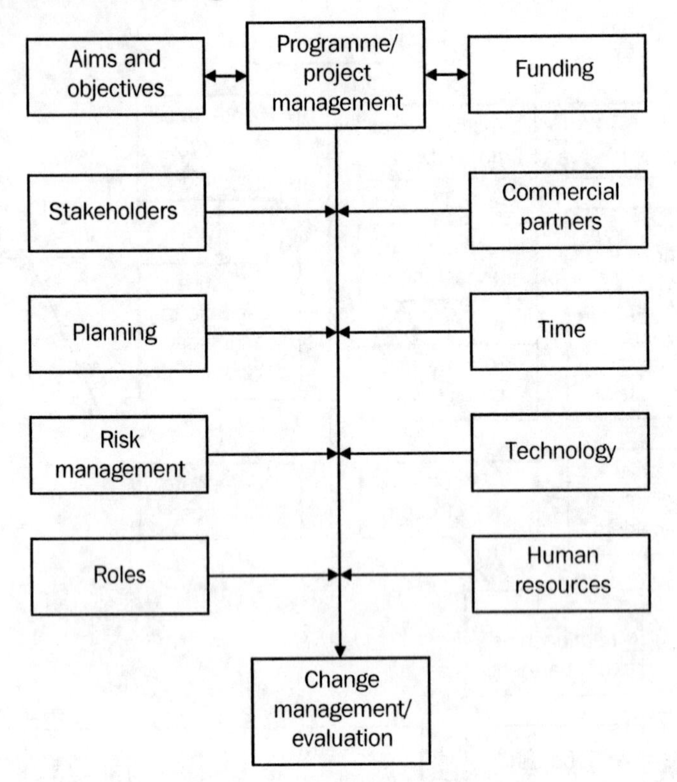

Figure 5.9 Programme management

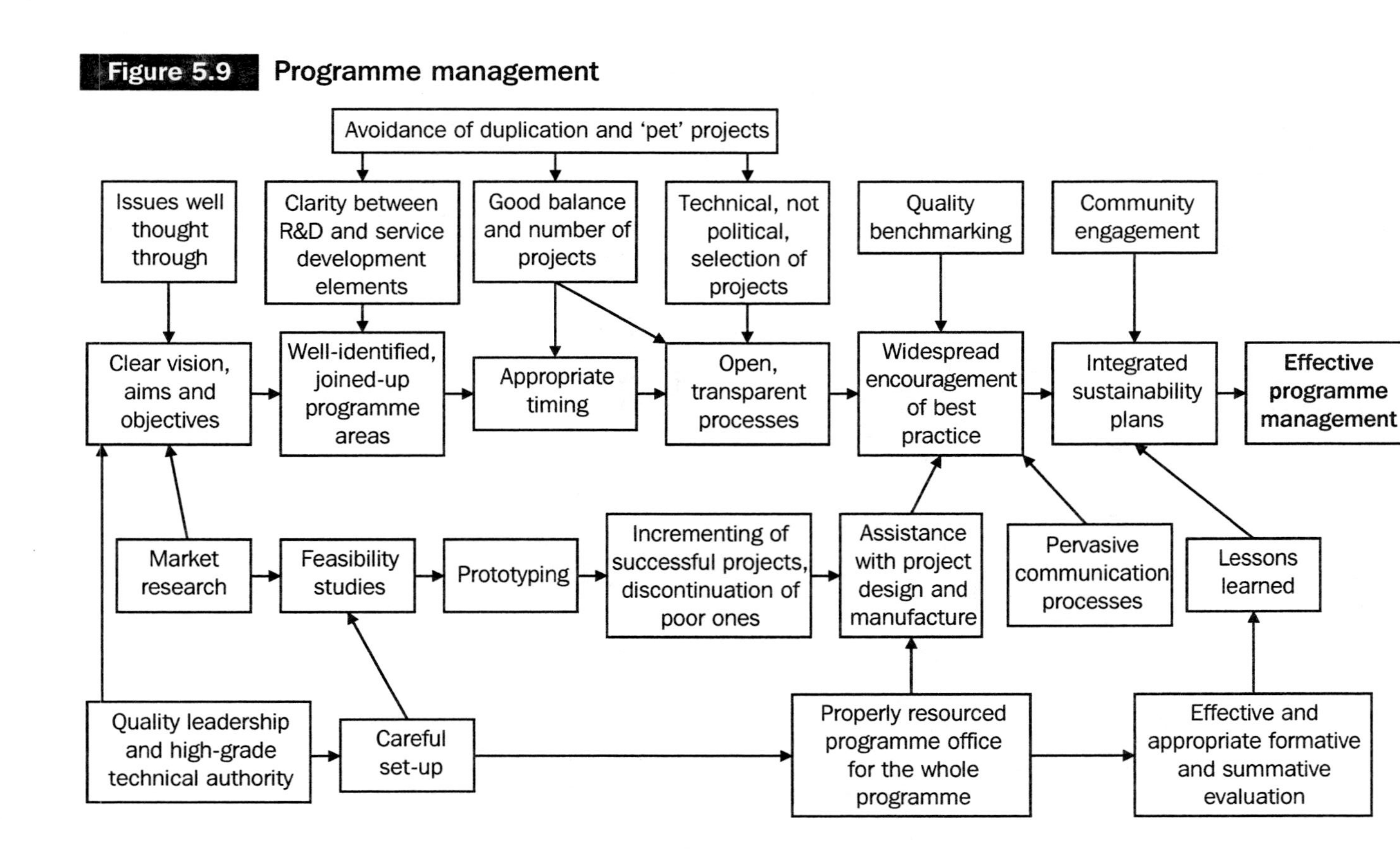

Figure 5.10 Project management

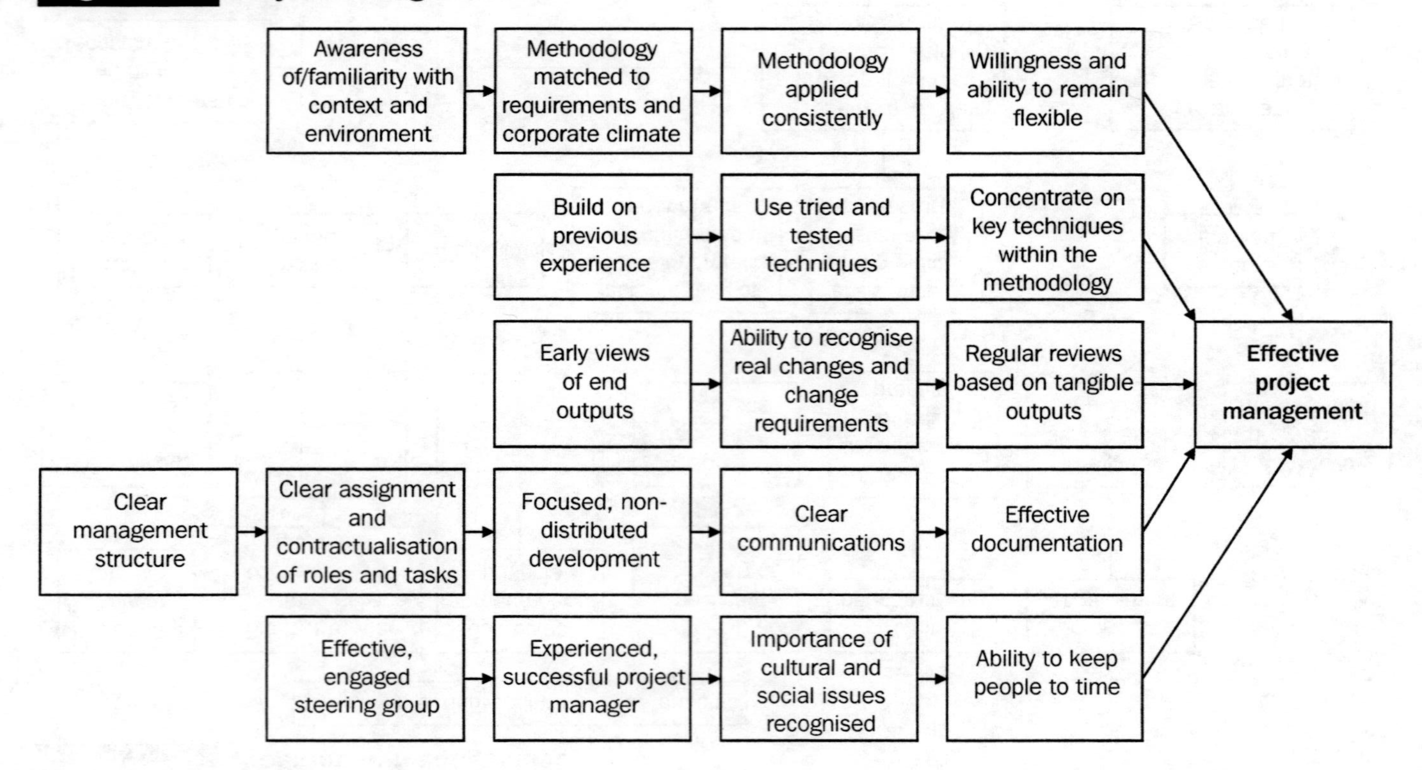

Figure 5.11 Aims and objectives

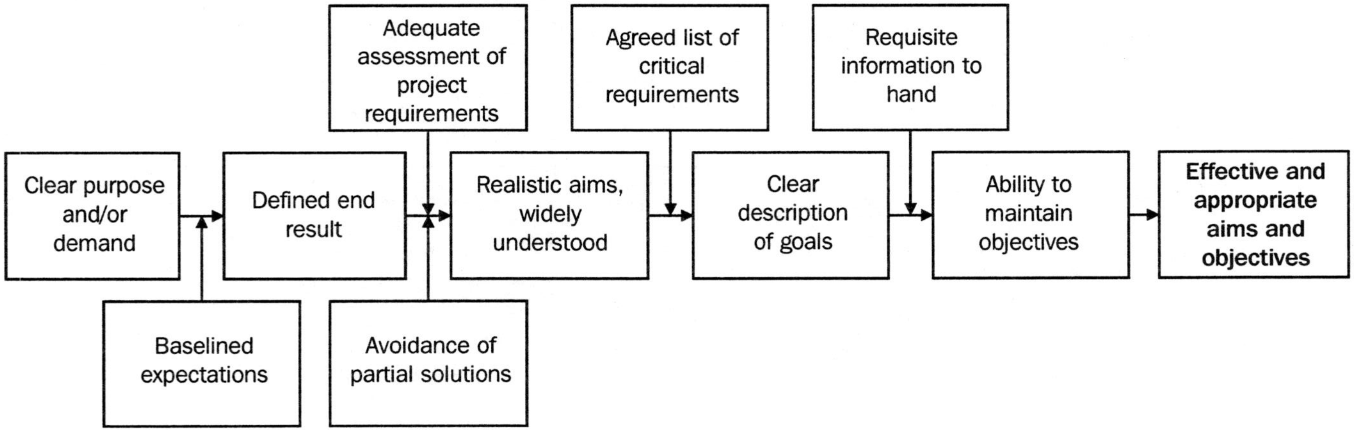

Figure 5.12 Funding

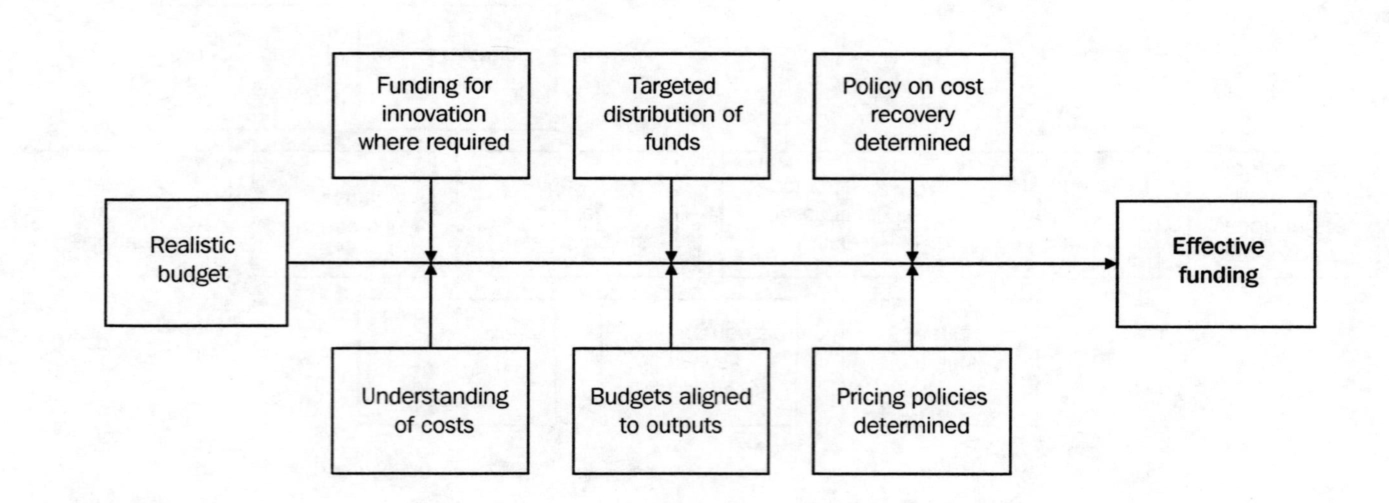

Figure 5.13 Stakeholders/commercial partners

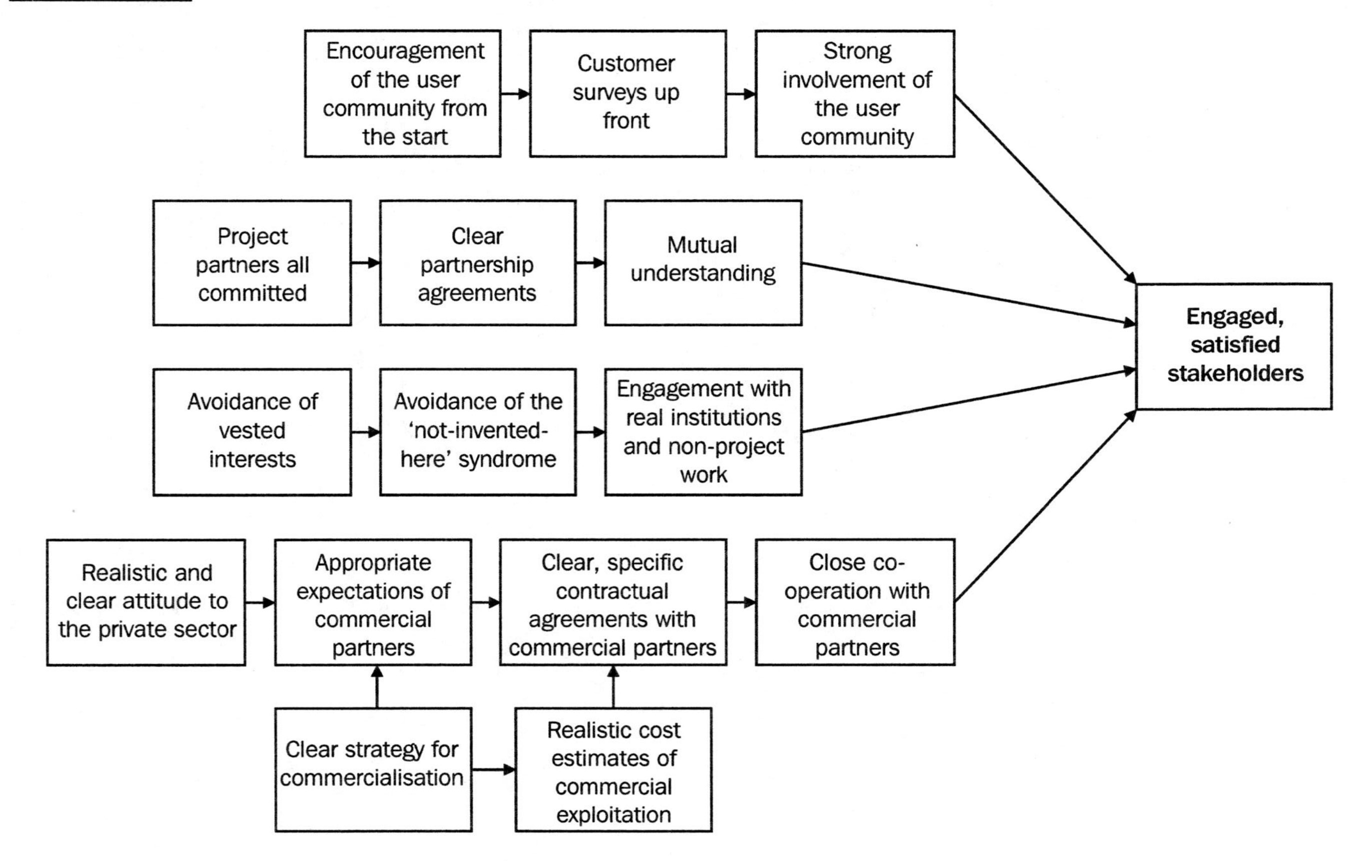

Figure 5.14 Planning/time

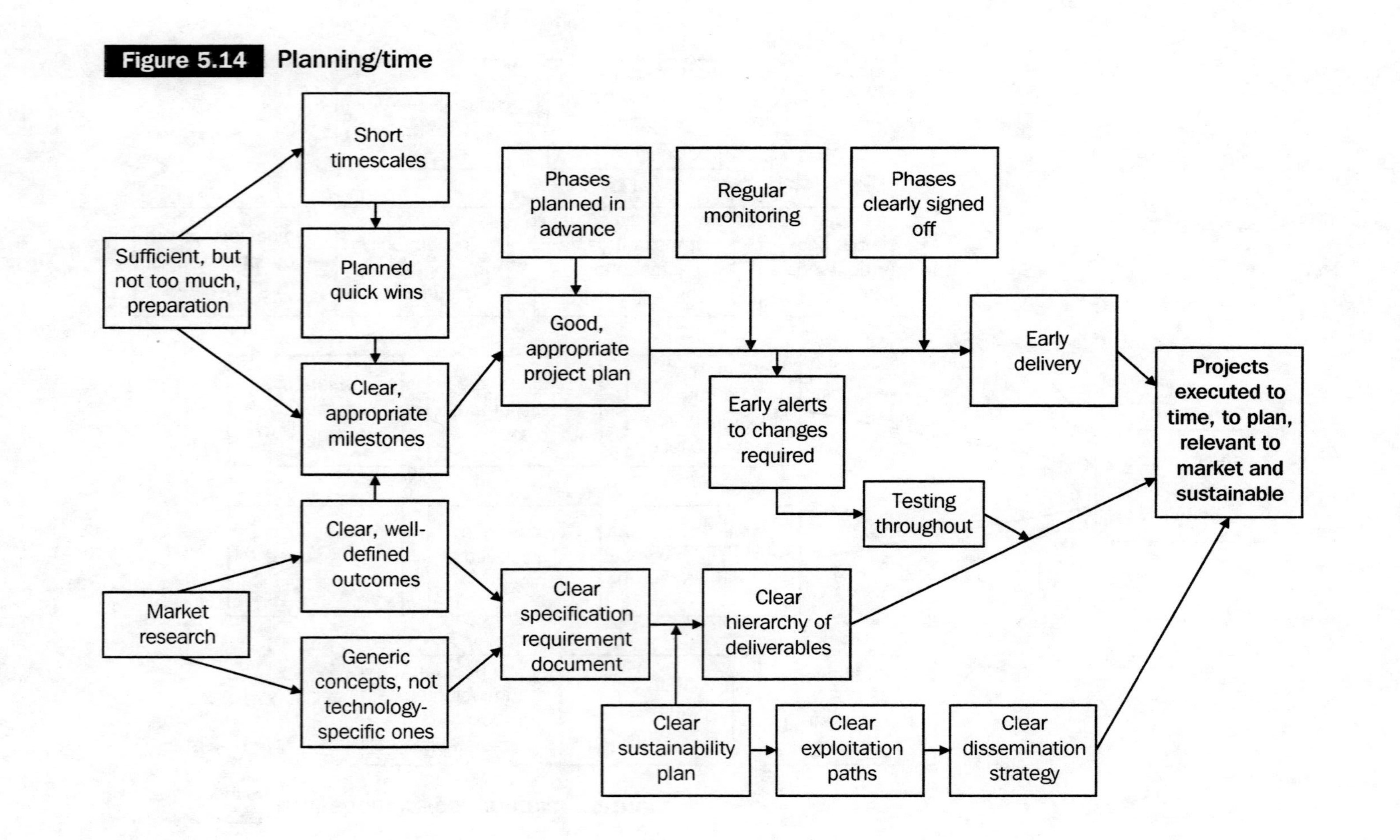

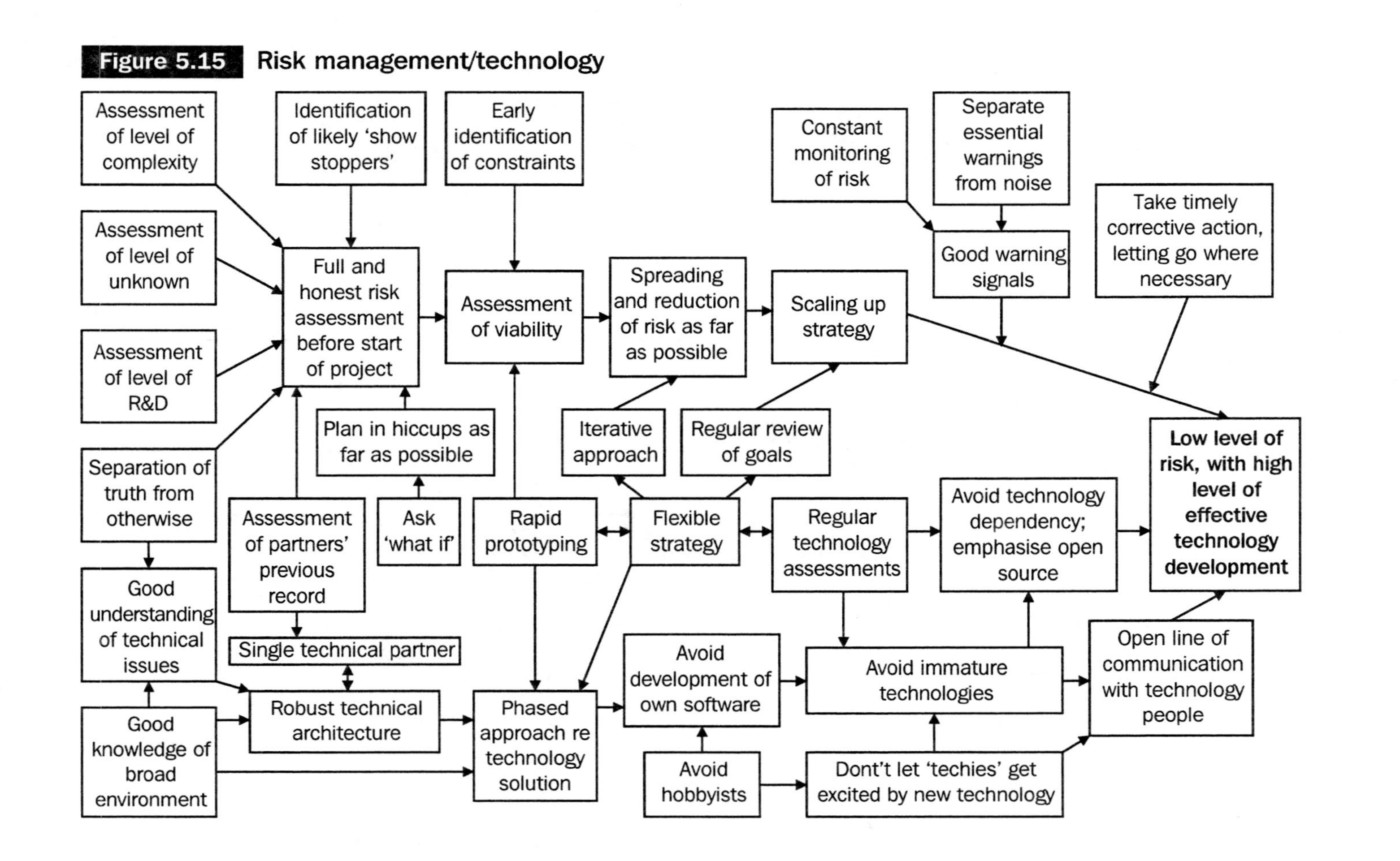
Figure 5.15 Risk management/technology
Assessment of level of complexity
Identification of likely 'show stoppers'
Early identification of constraints
Constant monitoring of risk
Separate essential warnings from noise
Assessment of level of unknown
Full and honest risk assessment before start of project
Assessment of viability
Spreading and reduction of risk as far as possible
Scaling up strategy
Good warning signals
Take timely corrective action, letting go where necessary
Assessment of level of R&D
Plan in hiccups as far as possible
Iterative approach
Regular review of goals
Low level of risk, with high level of effective technology development
Separation of truth from otherwise
Assessment of partners' previous record
Ask 'what if'
Rapid prototyping
Flexible strategy
Regular technology assessments
Avoid technology dependency; emphasise open source
Good understanding of technical issues
Single technical partner
Avoid development of own software
Avoid immature technologies
Open line of communication with technology people
Robust technical architecture
Phased approach re technology solution
Good knowledge of broad environment
Avoid hobbyists
Dont't let 'techies' get excited by new technology

Figure 5.16 Roles/human resources

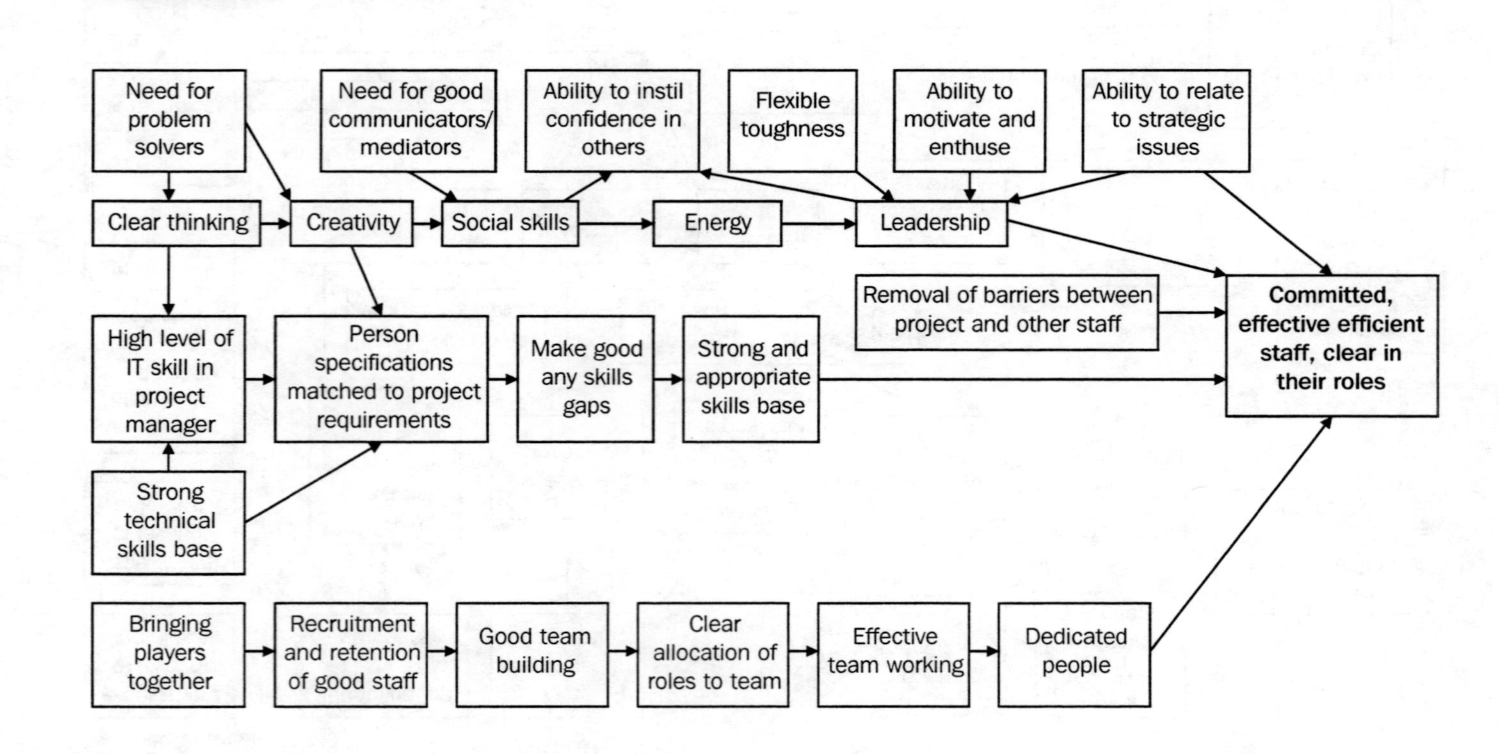

Figure 5.17 Change management/evaluation

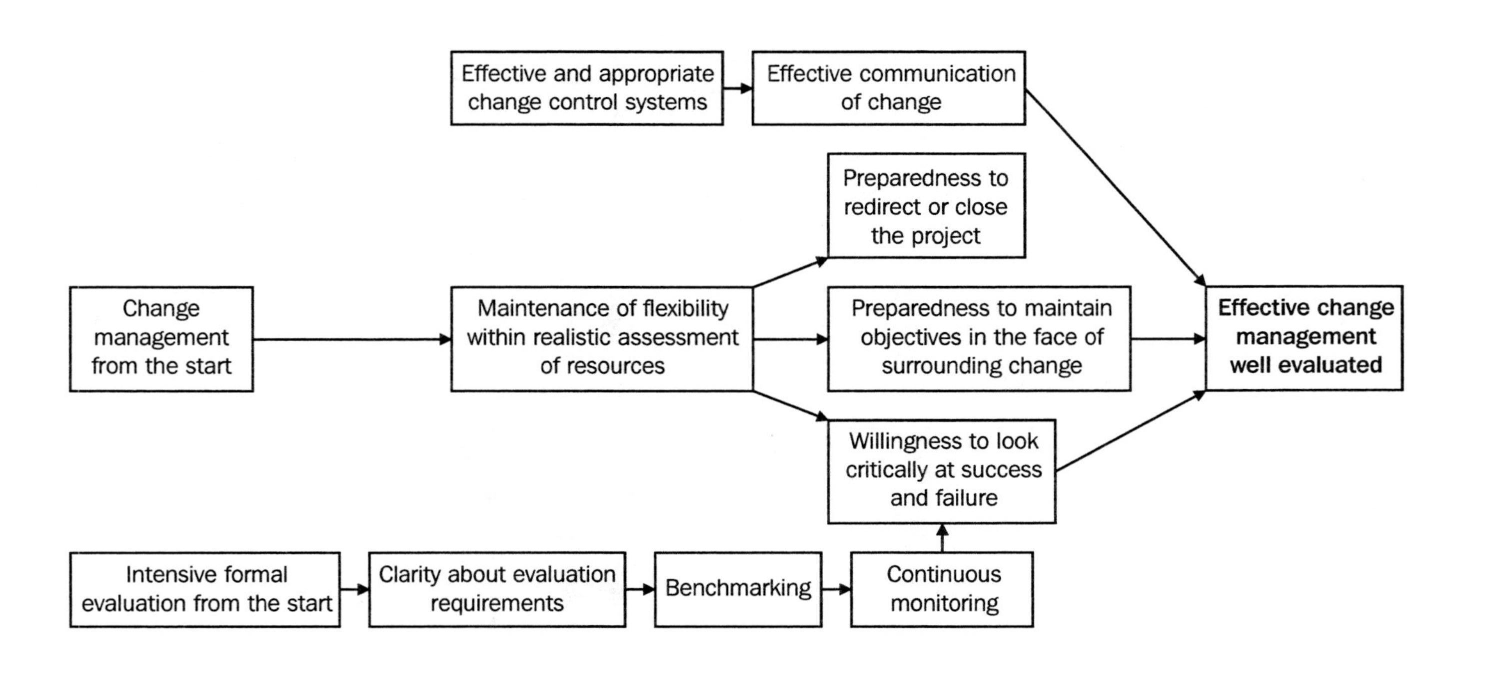

FMEA

Failure mode and effects analysis (FMEA) is a useful way of managing risk in new technology developments. It has a long history, being first used in the aerospace and automotive industries. Its application is now widespread in a wide range of environments. The technique's main aims are to:

- Anticipate the main problems.
- Identify where those problems are most likely to happen.
- Analyse and assess the effects of the problems identified.
- Facilitate the development of a plan to prevent the problems identified actually happening and/or to mimimise their effects if they do.

FMEA focuses on the requirements of those who will use the technology application, once available. At each stage of the process, therefore, it is important to keep asking: 'Who are the users who are to benefit from this project?' FMEA can be carried out over a whole programme, or parts of it, or in relation to one specific element, albeit as one tool in a whole collection of tools. The approach might best be commissioned by senior management as part of an overall quality improvement programme. However, given the general usefulness of FMEA, initiation could be by managers at many different levels. It is particularly useful for activities where risk and the effect of the proposed responses to that risk need in particular to be identified. High-risk results will point the way both to 'breakthrough' projects and to priority objectives for control. FMEA scores can also be used to assess the effectiveness of the teams involved. ('How quickly do they identify key risks and reduce them? How accurate are their analyses?)

An FMEA is typically carried out using an analysis sheet. This can be adapted to suit local conditions and specific projects. One model is given below, with the worked example of a new electronic document delivery service. Each stage of the proposed project is then analysed and data entered into the sheet, as described below. The more an FMEA is used in similar kinds of technology projects, the easier it is to predict where the problems will be. A significant advantage of the technique is that it can be used in a wide range of diverse situations. The key disadvantage is that it relies on both subjectivity (the scores and rankings are a matter of opinion rather than hard scientific fact) and historical experience ('this happened the last time we used this supplier').

The main aspects of FMEA are as follows:

1. Project stage

This should be a concise description of the stage of the project, as with, for example, 'analysis of alternative models'.

2. Potential failure mode

There are likely to be many ways in which the project fails to meets its aims. Each likely cause of failure is described as a 'failure mode'. In technology projects, one particular cause is often the inadequate or inappropriate skills of those undertaking the project. Each of these modes should be listed. Failure, in this context, is either a failure to deliver what was promised or the delivery of something that is unacceptable to, or inappropriate for, the user.

3. Potential effects of failure

It is then necessary to describe the potential effects of the failure mode on the project – including the next stages rather than the project overall – and the project's desired outcomes. If the potential effects are on the final outcomes or deliverables of the project, then it is important to consider the effects that these may have on the users. If the effects impact more particularly on the rest of the project, then it is necessary to consider the key stakeholders in the remaining stages and the requirements of that stage of the project if they are not likely to be met. In the case of a document delivery system, for example, it may be that the end-product is too cumbersome to use (e.g. it searches bibliographic databases sequentially rather than in parallel, wasting user time), or the technology developed for use in the new system does not integrate well with existing library housekeeping systems and requires time-consuming process changes and manual intervention.

4. Severity ranking (SR)

The SR measures the seriousness of the effect of the failure mode on the user. The level of severity is typically measured through a ranking process such as that shown in Table 5.1.

Table 5.1 Severity ranking

Level	Severity	Ranking
Minor	The user is unlikely to notice any problems. The failure can be rectified easily and without any real impact on either cost or quality.	1–2
Low	The user will notice the effect but is unlikely to be more than a little irritated by the change to expected outcomes. The project will be delayed, but only by a little and additional costs will be minimal.	3–4
Moderate	The user will be noticeably dissatisfied with the results of the project. Quality will be lower than expected and the original timetable for completion will be exceeded. Some aspects of the project may have to be re-done, causing extra expense.	5–6
High	The user will be severely dissatisfied with the results of the project. The introduction of the project as a working service or product may have to be deferred. It is highly likely that the project will be unsuccessful in terms of its original stated objectives and outcomes.	7–8
Very high	The user will be almost wholly dissatisfied with the results of the project. In fact, they may not see any of the desired outcomes and benefits that were promised at the outset of the project. The project is likely to be deemed a failure and may be cancelled before the end of the allotted time period.	9–10

5. Potential reasons for failure

The main reasons why the project or the stage failed are listed. There are likely to be many possibilities with a technology project. They include: lack of skills, poor project management, insufficient resources allocated, fragility of the technology, poor performance of the technology provider, etc.

6. Likelihood ranking (LR)

The LR measures the extent to which a potential cause of failure will actually happen. The assessment of likelihood should be carried out before any preventive action is carried out. As with the severity ranking, a 10-point scale is recommended (see Table 5.2).

Table 5.2 Likelihood ranking

Level	Likelihood	Ranking
Certain	Has occurred or will definitely occur.	10
Very high	Will almost certainly occur as an 'odds on favourite'.	8–9
High	Likely to occur (50 per cent chance).	6–7
Moderate	Could occur (less than 50 per cent but more than 10 per cent chance)	4–5
Unlikely	Unlikely to occur (1 in 10 to 1 in 100)	2–3
Extremely unlikely	Highly unlikely to occur (well over 1 in 100)	1

7. Priority risk factor (PRF)

The PRF is expressed as a number. It is calculated by multiplying the SR and LR numbers. It provides a good indication of relative priorities in terms of managing risk.

8. Prevention plan (PP)

The PP should include a description of all the actions that need to be undertaken in order to minimise the failure modes identified.

9. Plan effectiveness number (PEN)

The PEN is a rating of the likely effectiveness of the preventive actions. The PEN is calculated using a ranking grid such as that shown in Table 5.3.

10. Residual risk factor (RRF)

The RRF is the final and most important calculation. It is obtained by multiplying the PRF by the PEN. The result provides managers with a numerical value that sums up the level of risk that remains after all the preventive measures have been put in place. If the RRF is too high, then further ways will have to be found of minimising risk or an alternative strategy will have to be devised. In projects with a large research and development component, the high level of residual risk is justifiable, provided that lessons can be learned from the project, even if it fails.

Table 5.4 shows a partial worked example concerning a (fictitious) project to develop a new electronic document delivery service. The FMEA assumes that the technology to be adopted is relatively new and yet to be tried and tested in any major LIS environment. The analysis sheet shows that the key area of remaining risk relates to the technology supplier. Is the RRF low enough to proceed? What additional prevention initiatives could be put in place? Should an alternative strategy be developed and adopted? If so, what is it? If the project is to be instigated or continued, then the FMEA can again be used to assess the relative risk levels once a further round of analysis has been conducted and further or enhanced prevention measures put in place. If a different project is to be contemplated, then the FMEA can be a useful tool in assisting option appraisal. It may be that the supposedly 'safer' alternative is just as risky as the original project.

Table 5.3 Plan effectiveness number

Level	Degree	Ranking
Certain	The PP will be certain to prevent failure in the areas identified	0
Extremely unlikely	If the PP is undertaken successfully, then the failure will not occur	0.1–0.2
Unlikely	It is unlikely that the failure will occur if the PP is carried out successfully	0.3–0.4
Moderate	Even if the PP is undertaken successfully, the failure may still occur	0.5–0.6
Likely	Even if the PP is undertaken successfully, the failure is likely to occur	0.7–0.8
Extremely likely	The failure is almost certain to occur regardless of the PP	0.9–1.0

Table 5.4 Failure mode and effects analysis: a worked example

Project stage/ process	Potential failure	Potential effects	SR	Potential cause of failure	LR	PRF	Prevention plan	PEN	RRF
General	Insufficient time devoted to the project; too many interruptions; changing nature of the technology	Inability to complete the project successfully/to a high standard	10	Lack of prioritisation; other, more pressing work demands within the organisation that have to take precedence	2	20	Develop a full plan for what should be done and when; allocate discrete resources to the project; identify a single project leader who will be able to devote sufficient time to the work; make sure that the senior management is fully supportive of the project and that there are good two-way communications between them and the project leader/ team	0.1	2

Analysis of alternative models	Inadequate/ inappropriate/ incomplete analysis	Insufficiently robust base on which to carry out an effective analysis and assessment for the project; inadequate contextualising of the findings	7	Insufficient time devoted to analysing the models; inexperience of the field; unavailability of sufficiently robust data regarding the alternatives	5	35	Plan time effectively and devote regular slots throughout the earlier stages of the project; take advice from library/ information technology specialists	0.1	3.5
Project definition	Aims and objectives defined too broadly	Project too large to complete in the time available	8	Insufficient thought given to the project and insufficient discipline in focusing on what is possible in the time	3	24	Continuous iteration of the topic; feedback from experts and the project board in response to reports from the project management; checking out project as a whole at this stage; anticipation of the final results	0.1	2.4

Table 5.4 Continued

Project stage/ process	Potential failure	Potential effects	SR	Potential cause of failure	LR	PRF	Prevention plan	PEN	RRF
Project design	Badly designed methodology	Outcomes inadequate, inappropriate or otherwise insufficiently robust	7	Insufficient time given to designing the methodology; inappropriate anticipation of the outcomes; inexperience at designing technology projects	3	21	Good time management; regular checking back with tutor and others about the appropriateness of the design; use of experts within the university to help; feedback from project board in response to reports from project management	0.2	4.2
Project commencement	Incomplete or inadequate programme of work	As above	8	Insufficient time given to undertaking the project; poor design of the methodology; inappropriate anticipation of the results; inexperience at designing technology projects	4	32	All the above actions	0.2	6.4

Supplier involvement	Inability of technology supplier to participate appropriately and effectively in the project	Incomplete, inappropriate or ineffective software; inability to implement a full service as originally envisaged	10	The supplier may not be sufficiently competent in the field; the supplier may have other customers to please at the same time or may have a different agenda for the development of the software	5	50	Clear articulation of the expectations of the supplier and the deliverables required, with dates and quality definitions contractually agreed between the parties	0.4	20
User satisfaction	Lack of take-up of the new service	Insufficient benefits from the new software and the associated service; lack of attractiveness of the outcomes of the project in terms both of benefits (e.g. more effective working) or the satisfaction levels in actually using the product (is it user friendly enough?)	10	The product design may be flawed; the data on user requirements might be insufficiently robust; the analysis of need and process may be inappropriate, because of inexperience or inappropriate anticipation in this area of work	3	30	Continuous iteration of the analysis; feedback from project board in relation to users (users need to be involved in the board and their views taken into account at all stages in the project); clear and continuous communication between the project management and the supplier at all stages; advice from independent experts in the field	0.2	6

Table 5.4 Continued

Project stage/ process	Potential failure	Potential effects	SR	Potential cause of failure	LR	PRF	Prevention plan	PEN	RRF
Project outcomes	Lack of usable product; inability to introduce the new service	Inadequate services and software; failure of project	10	Inadequate definition of project requirements; poor software design and implementation	3	30	See earlier comments	0.2	6

Summary

On the basis of this chapter, a clear model of success in IT project management emerges, based on a template of key attributes, drivers and risk management tools. A successful project is one where, above all:

- objectives are achieved;
- output is of the right quality;
- output is useful;
- the budget is met;
- the outcomes are flexible and adaptable and the stakeholders have the ability to re-purpose them.

Success is best achieved when there is/are at least:

- careful planning;
- realistic project aims;
- flexibility in application;
- excellent communications;
- an effective board or steering group.

Risk will be significantly reduced if there is/are a minimum of:

- high-quality management;
- appropriate staff selection;
- well-articulated milestones;
- commonly agreed deliverables;
- clear project plans.

A template that measures and forecasts success should contain the following elements:

- aims and objectives
- funding

- stakeholders
- commercial partners
- planning
- time
- risk management
- technology
- roles
- human resources
- change management
- evaluation
- programme and project management.

Any programme or project management must also be aware of the availability of resources and resource constraints, the appropriateness of methodologies used, technology issues and the overall environment in which the project is set. The ways in which these background factors are addressed will measure the effectiveness of the project or programme management.

Of particular importance in the context of IT development projects is the need to maintain strategic direction through an effective response to technology developments, supported by trend analyses and the identification and maintenance of an appropriate skills base. There is a need to avoid both 'technocratic utopianism' (Davenport, 1994) and 'pet projects', hobbyists or partial solutions.

The need for realistic budgets based on an understanding of costs, targeted distribution of funds and allocations aligned to outputs was stressed. Programme management should determine the extent to which the objective is to produce income-generating services and, if so, determine the policy for income generation/cost recovery. Innovation needs resources, and this also must be recognised.

Success and the degree of risk are contingent upon the amount of R&D and the level of 'unknown' within the project. An assessment of these aspects is an important element of both risk management and likely success.

The need for realistic agreements and good communications with commercial partners (preferably as few as possible) is of paramount importance where technology development work is involved. So too is the involvement of user communities from an early stage, not least through market research. Early prototyping, short timescales, incrementing of successful projects and closure of unsuccessful ones also appeared as key responses to the 'moving goalposts' of technology development. They should be supported by early views of end outputs (based on clear requirements specifications), continuous monitoring of progress, regular testing of outputs, early alerts to required changes and baselined expectations.

Fifty per cent of project management is about people. Whereas the human resource elements of IT project management are generic ones relating to team building, recruitment of good staff and clear allocation of roles to the team, there are specific issues relating to IT development projects. These concern the need for a strong and appropriate skills base, with a high level of skill in the project manager who can provide both quality leadership and high-grade technical authority. This will help projects to separate out the truth. Cultural and social issues, as recognised by Gray (2001), are also important. They range from the crucial importance of the team dynamic, through the removal of barriers between project and other staff to pervasive communication processes.

In conclusion:

- Managers should use a clear model of success in IT project management to assist in the formulation of programmes and projects.

- Actual and potential management effectiveness has to be assessed.
- The greater the degree of new work, the higher the risk and the greater the need for tight management.
- Particular attention should be paid in the formulation and management of IT development projects to the maintenance of strategic direction, avoidance of pet or hobby projects and realistic budgeting (especially where R&D or innovatory implementation is involved).
- Long projects should be avoided. Early views of end outputs based on clear, user-based requirements, short timescales, prototyping, early delivery of results and incrementing of successful projects will assist management to cope with the shifting goalposts of IT.
- There should be realistic agreements with commercial suppliers. The fewer the number of suppliers, the greater the chance of success.
- There should always be an appropriate skills base in any IT project. The technical authority of the project manager will be crucial to success. Project managers will also need to be able to communicate well and build teams. General management will need to integrate project and mainstream workers as far as possible.

Bibliography

Ambler, S.W. (2001) 'Planning modern day software projects', *Computing Canada*, 27(4), 11–16.

Artto, K.A. et al. (2001) 'Managing projects front-end: incorporating a strategic early view to project management with simulation', *International Journal of Project Management*, 19(5), 255–65.

Atkinson, R. (1999) 'Project management: cost, time and quality, two best guesses and a phenomenon; it's time to accept other criteria', *International Journal of Project Management*, 17(6), 337–42.

Baccarini, D. and Archer, R. (2001) 'The risk ranking of projects: a methodology', *International Journal of Project Management*, 19(3), 139–46.

Bradley, K. (1997) *Understanding PRINCE 2*. Bournemouth: SPOCE Project Management Ltd.

Chapman, C. (1997) 'Project risk analysis and management – PRAM the generic process', *International Journal of Project Management*, 15(5), 273–81.

Clarke, A. (1999) 'A practical use of key success factors to improve the effectiveness of project management', *International Journal of Project Management*, 17(3), 139–45.

Cooper, R.G. and Kleinschmidt, E.J. (1987) 'New products; what separates winners from losers?', *Journal of Production Innovation Management*, 4, 169–84.

Davenport, T.H. (1994) 'Saving IT's soul: human-centered information management', *Harvard Business Review*, March–April, 119–31.

Davis, J. et al. (2001) 'Determining a project's probability of success', *Research Technology Management*, 44(3), 51–8.

Dey, P.K. (1999) 'Process re-engineering for effective implementation of projects', *International Journal of Project Management*, 17(3), 147–59.

Drummond, H. (1999) 'Are we any closer to the end? Escalation and the case of Taurus', *International Journal of Project Management*, 17(1), 11–16.

ESYS Consulting (2001) *Summative Evaluation of Phase 3 of the E-lib Initiative: Final Report Summary*. London: ESYS Consulting.

Feldman, J.I. (2001) 'Project recovery: saving troubled projects', *Information Strategy: The Executive's Journal*, 17(2), 6–12.

Fowler, A. and Walsh, M. (1999) 'Conflicting perceptions of success in an information systems project', *International Journal of Project Management*, 17(1), 1–10.

Gray, R. (2001) 'Organisational climate and project success', *International Journal of Project Management*, 19(2), 103–10.

Heeks, R., Mundy, D. and Salazar, A. (1999) *Why Health Care Information Systems Succeed or Fail*. Manchester: University of Manchester Institute for Development Policy and Management.

Hvam, L. and Have, U. (1998) 'Re-engineering the specification process', *Business Process Management Journal*, 4(1), 25–43.

Jaafari, A. (2000) 'Life-cycle project management: a proposed theoretical model for development and implementation of capital projects', *Project Management Journal*, 31(1), 44–53.

Jaafari, A. (2001) 'Management of risks, uncertainties and opportunities on projects: time for a fundamental shift', *International Journal of Project Management*, 19, 89–101.

Jelinek, M. and Schoonhoven, C.B. (1990) *The Innovation Marathon: Lessons from High Technology Firms*. Oxford: Blackwell.

Jiang, J.J. (2000) 'Project risk impact on software development team performance', *Project Management Journal*, 31(4), 19–27.

Johnson, J. et al. (2001) 'The criteria for success', *Software Magazine*, 21(1), 1–8.

JISC (Joint Information Systems Committee) (2002) *Circular 1/02: Focus on Access to Institutional Resources Programme*. Bristol: JISC.

Kirby, E.G. (1996) 'The importance of recognizing alternative perspectives: an analysis of a failed project', *International Journal of Project Management*, 14(4), 209–11.

Kumar, N. et al. (2000) 'From market driven to market driving', *European Management Journal*, 18(2), 129–42.

Kuprenas, J.A. (2000) 'Project manager workload – assessment of values and influences', *Project Management Journal*, 31(4), 44–52.

Lewis, A.C. (1995) 'The use of PRINCE project management methodology in choosing a new library system at the University of Wales Bangor', *Program*, 29(3), 231–40.

Lopes, M.D.S. and Flavell, R. (1998) 'Project appraisal – a framework to assess non-financial aspects of projects during the project life cycle', *International Journal of Project Management*, 16(4), 223–33.

McDonald, J. (2001) 'Why is software project management difficult? And what that implies for teaching software project management?', *Computer Science Education*, 11(1), 55–71.

Munns, A.K. and Bjeirmi, B.F. (1996) 'The role of project management in achieving success', *International Journal of Project Management*, 14(2), 81–7.

Nellore, R. and Balachandra, R. (2001) 'Factors influencing success in integrated product development (IPD) projects', *IEEE Transactions on Engineering Management*, 48(2), 164–75.

Noori, H. (1990) *Managing the Dynamics of New Technology: Issues in Manufacturing Management*. Englewood Cliffs, NJ: Prentice Hall.

Orwig, R.A. and Brennan, L.L. (2000) 'An integrated view of project and quality management for project-based organisations', *International Journal of Quality and Reliability Management*, 17(4), 351–63.

Pender, S. (2001) 'Managing incomplete knowledge: why risk management is not sufficient', *International Journal of Project Management*, 19, 79–87.

Pinto, J.K. (2000) 'Understanding the role of politics in successful project management', *International Journal of Project Management*, 18, 85–91.

Raz, T. and Michael, E. (2001) 'Use and benefits of tools for project risk management', *International Journal of Project Management*, 19, 9–17.

Roussel, P.A. et al. (1991) *Third Generation R&D: Managing the Link to Corporate Strategy*. Boston: Harvard Business School.

Sampath, R.S.V. (2001) 'PSM standards for effective project management', *Hydrocarbon Processing*, 80(5), 92-A.

Stewart, W.E. (2001) 'Balanced scorecard for projects', *Project Management Journal*, 32(1), 38–54.

Sundbo, J. (1997) 'Management of innovation in services', *Service Industries Journal*, 17(3), 432–55.

Tavistock Institute (2000) *1999 Synthesis of Elib Annual Reports: Phase 2 and Phase 3*. London: Tavistock Institute.

Taylor, W.J. and Watling, T.F. (1973) *Practical Project Management*. London: Business Books.

Thamhain, H.J. (1990) 'Project management in the factory', in D.I. Cleland and B. Bidanda (eds), *The Automated Factory Handbook: Technology and Management*. Blue Ridge, PA: TAB Books.

Thite, M. (2000) 'Leadership styles in information technology projects', *International Journal of Project Management*, 18, 235–41.

Thomas, S.R. (1999) 'Compass: an assessment tool for improving project team communications', *Project Management Journal*, 30(4), 15–25.

Thoms, P. and Pinto, J.K. (1999) 'Project leadership: a question of timing', *Project Management Journal*, 30(1), 19–27.

Twiss, B. (1992) *Managing Technological Innovation*. Harlow: Longman.

Twiss, B. and Goodridge, M. (1989) *Managing Technology for Competitive Advantage*. London: Pitman.

Uher, T.E. and Toakley, A.R. (1999) 'Risk management in the conceptual phase of a project', *International Journal of Project Management*, 17(3), 161–9.

Vadapalli, A. and Mone, M. (2000) 'Information technology project outcomes: user participation structures and the impact of organisation behaviour and human resource management issues', *Journal of Engineering Technology Management*, 17, 127–51.

Vandersluis, C. (2001) 'Almost all pilot projects lack measuring metrics', *Computing Canada*, 27(11), 13–18.

Ward, S. (1999) 'Requirements for an effective project risk management process', *Project Management Journal*, 30(3), 37–44.

Webster, G. (1999) 'Project definition – the missing link', *Industrial and Commercial Training*, 31(6), 240–5.

Williams, T.M. (1997) 'Empowerment vs risk management?', *International Journal of Project Management*, 15(4), 219–22.

Zwikael, O. et al. (2000) 'Evaluation of models for forecasting the final cost of a project', *Project Management Journal*, 31(1), 53–8.

6

Summary

Components of strategic technology management

Johnson and Scholes (1993) sum up the strategic process as consisting of: analysis, choice and implementation. Strategic technology management has a number of components. Environmental analysis is a prerequisite of strategy formulation. Chapter 1 included a description of a strategic IQ test that evaluated the extent to which an organisation was capable of developing and implementing a strategy. Betz (1993) would describe this as a 'strategic attitude'. Others might call it a learning organisation, as also discussed in Chapter 1. A strategic attitude without a strategic plan will be of limited use, just as a plan without the necessary organisational capacity will be of relatively little value.

The process should consider the expectations and requirements of the various stakeholders and the culture of the organisation. External and internal strengths, weaknesses, opportunities and threats (SWOT) will have to be identified. In particular, the organisation's readiness to develop and implement a technology strategy and its willingness and ability to embrace change will need to be addressed. All this work is done in the context of forecasting the future and determining the strategic position that the organisation can realistically take to best advantage. Delphi and competitive or strategic web techniques can be especially useful here. But strategy is also about solving specific problems and various types of systems thinking can be used to good effect to underpin the effectiveness of a strategy's implementation. Ultimately, a number of strategic options need to be identified and a choice made as to the preferred option for implementation. This book has also stressed the importance of time and timing. Time is required to develop new products and services and good timing is needed in order to launch the most appropriate technology onto the market in the most hospitable environment.

A strategy is meaningless unless it is implemented. Implementation is likely to involve change management and a structured approach is recommended for technology strategy implementation. Chapter 5 dealt with the questions of programme and project management and success and failure in such work. The longer the time frame for the strategy, of course, the greater the degree of uncertainty, in terms of both the technology and the underlying environment in which it will be introduced and operated:

> Strategy applies in different ways over different time horizons. Short-term projects are more amenable to conventional strategic planning, while longer-term aims are broader, a question of corporate direction. Both are a balance between planned and unpredictable development but to different extents. The distinction is also between small scale, containable projects and complex, unpredictable programmes with potentially critical impact. (Grindley, 1993)

Uncertainty brings risk and risk management is an essential element of strategy implementation. Effective programme and project management is about taking and implementing good decisions and the ways in which this can happen has been explored. Good-quality resource allocation should be one of the benefits of a good strategy. Taking a strategic view will ensure that resources are diverted to or from the areas where they are most or least needed over a period of time that is sufficiently long for benefits to be identified and yielded. Strategy relates not only to implementation but also to operation.

Integrative approaches

Good strategic technology management requires the integration of any technology strategy with the rest of an organisation's strategies. These may be the business or marketing strategies, the research and development strategy and the human resource strategy. As Anstey (2000) has noted, the people elements of technology development and application are likely to be of paramount importance. Sethi et al. (1985) summarise the integration of a technology strategy with the rest of an organisation's strategic planning as follows:

> The purpose of business planning is to determine the most profitable manner of allocating limited resources among competing alternative profit opportunities. The technique for technology strategies' insertion into the overall planning starts by defining the [organisation's] technology profile. This is an overall assessment of current products [or services], R&D investment and organisation. Next, an internal and external scan of [the] technology environment is undertaken, which goes beyond the limits of the traditional business cycle. It is followed by technology directions and incorporates these effects into future scenarios, to enable a complete evaluation of the [organisation's strategy] in a time frame considering other factors, such as economic and environmental ones.
>
> This exercise sets the stage for identifying and systematically analyzing key corporate technology alternatives and technology priorities. This is followed by integrating the

previously developed technology portfolio into the overall corporate strategy, thus assuring consistent objectives and effective implementation. During the last stage, technology investment opportunities are determined.

Strategy in practice

The techniques of strategic technology management discussed thus far represent 'means to an end' and not 'ends' in themselves. As several writers (e.g. Quinn, 1988; Mintzberg, 1994) have pointed out, strategy development can improve the chances of success, but they cannot guarantee it. Quinn in particular considers that there is a continuous need to increment the strategy by the acquisition of additional information over time and the consequent reduction of risk. Because strategy tries to respond to the unknowable as well as the unpredictable, it is essential that the strategic positioning of the organisation that results from strategy development is sufficiently robust and flexible enough for the stated aims and objectives to be achieved regardless of changes in either the technology or the surrounding environments, even where these changes were not – and could not be – forecast.

It is highly unlikely that a strategy will be implemented exactly as intended. As Johnson and Scholes (1993) suggest, the 'realised strategy' may be the result of what actually emerges from the implementation of the strategy, and which takes account of any threats or opportunities not originally intended and which responds to third-party interventions, such as those made by governments. It is also important to note that organisations – and individuals working in them – may say one thing and do another. In other words, the organisation may think that it is following a strategy because the management says that it is when, in reality, it is following a different one. The theory that is 'espoused' within the organisation is not the one that is 'put into use'. The management will have to ensure that it is not deluding itself into believing that all is well. In addition, strategic drift may occur over time. This is where the strategy falters or stalls, requiring either revision or renewal and strong senior management input.

At various points in this book the need for flexibility has been stressed. Without an ability to adapt and change – rapidly and effectively – an organisation is likely to be left badly behind in the fast moving world of technology and technology management. The genuine variation of a strategy to take account of changed circumstances, technologies and environments should be seen as a sign of strategic and managerial strength rather than weakness.

Although a strategy should be about vision, it is important to remember that a number of constraints will be placed upon the strategic technology manager in its implementation. There is no completely free choice in strategic positioning, and environmental factors and events are likely to condition if not limit an organisation's ability to anticipate or respond to trends and opportunities. A strategy might propose 'fast follower' status, for example, and the reality might be that the organisation ends up in the lead, whether it likes it or not. This may be the case, for example, with next generation systems where a number of LIS units are all in contention to be the first to sign a contract with an innovative supplier.

Technological innovation and change is complex and characterised by a high degree of risk, typically brought about by uncertainty. The effective management of technological innovation requires a deep and sophisticated understanding of the many variables involved, together with the complex interactions between them, the overall environment and the stakeholders involved in the innovation process. It is therefore of crucial importance that an organisation is properly positioned to take real advantage of innovations.

Endnote

This book has been concerned with strategic technology management. It began with a discussion of strategy and management and their importance in the context of technology development, implementation and operation. Strategy development is a challenging but potentially rewarding task. The result should be a framework that facilitates the taking of decisions both in the short and the longer term:

> ... [its value] is that it implies that technological change should be seen not as a separate activity to be studied in its own terms, but as integral to [an organisation's] operations ... technology must be seen as part of business ... Strategy immediately includes the notions that, first, the purpose of indulging in technological change is to contribute to overall ... objectives ... Second, strategy considers the interaction of the [organisation] with its environment, both with competitors and with outside sources of technology, and depends on the general rate of technical progress. Third, it is bound up with the internal organisation process, how organisation affects technological choice, implementation, and evaluation, and conversely how technology may itself bring about changes in organisation. (Grindley, 1993)

Bibliography

Anstey, P. (2000) *C & IT Skills: Developing Staff C & IT Capability in Higher Education*. Norwich: University of East Anglia.

Betz, F. (1993) *Strategic Technology Management*. New York: McGraw-Hill.

Grindley, P. (1993) 'Firm strategy and successful technological change', in A. Cozijnsen and W. Vrakking (eds), *Handbook of Technical Management*. Oxford: Oxford University Press.

Johnson, G. and Scholes, K. (1993) *Exploring Corporate Strategy*. Englewood Cliffs, NJ: Prentice Hall.

Mintzberg, H. (1994) *The Rise and Fall of Strategic Planning*. Englewood Cliffs, NJ: Prentice Hall.

Quinn, J.B. (1988) 'Managing innovation: controlled chaos', in J.B. Quinn et al. (eds), *The Strategy Process*. Englewood Cliffs, NJ: Prentice-Hall.

Sethi, N.K. et al. (1985) 'Can technology be managed strategically?', *Long Range Planning,* 18(4), 89–99.

The case studies

Technology in a changing world: the future of LIS work

Introduction

Four case studies have been included in this book as a means of providing exemplars of the themes, issues and approaches that have been described. The case studies are preceded by the following contextual commentary.

The 'information society' and LIS work

In recent years, there has been much talk about the 'information society':

> Politicians, those in the media and (sometimes) academics will use it as a 'buzz word' to describe the impact of information and communication technologies. The information society is much more than that: it can be a set of theoretical perspectives outlining the recent changes in society, it can be used to analyse different scenarios for current and future developments in society, and it can be used by information professionals to understand the impact that these changes are having on their

> role and the changing needs of their users. (Hornby and Clarke, 2003)

As Feather (2003) points out:

> The growth of some of the key areas of the information sector of the economy in the 1980s and 1990s is associated with technological innovation and the widespread adoption of new technologies ... The issue is whether this change is a fundamental re-conceptualisation of activities and their purpose, or whether it is merely a change, albeit very significant, in techniques and mechanisms. Another way of expressing the problem is to ask whether technological change drives social change, or social change demands new technological solutions to new and newly identified problems.

He continues:

> The issue is not whether there has been technological change and innovation: that is undeniable. It is rather whether the change has been led by the needs of users and of society at large, or whether technologies and systems have been developed for which uses have then been found ... [Some] writers have identified a consistent delay between the development of information and communication technologies and their large-scale adoption ... It is argued that there is a consistent pattern of invention followed by a search for an application. The application often turns out to be quite different from that envisaged by the inventor ...

Financial pressures

Traditional library services have increasingly been under pressure 'in the face both of increased costs and decreased funding support.

Many predict that migration to electronic services is a way to escape this negative feedback loop, as the digital path facilitates resource sharing, which should cut costs and improve customisation, which in turn will contribute to service value ... Within the budget paradigm itself, there are difficulties. Current practice and the traditional division of budgets into operating, capital and personnel are increasingly difficult to sustain' (Snyder and Davenport, 1997). 'Libraries are being asked to perform what sounds like a magic trick. They are to downsize, economize, and streamline, while at the same time improve quality and provide customers with services they value, and, as if these challenges are not enough, libraries are in the midst of a fundamental transformation brought about by technology' (Robinson and Robinson, 1994).

LIS departments

LIS departments are typically 'service industries'. They do not manufacture products; rather they provide a service or services that normally use technology rather than create it. Murdick et al. (1990) note that technology management in service industries is typically more challenging than in sectors where the output is normally a product. Firstly, the services provided will normally produce intangible rather than tangible outputs. Secondly, there is arguably a more intimate and integrated relationship between the user or customer in service rather than straight product-based industries, with the user actually participating in the process of service provision, for example through formal service level agreements or the creation of the input (such as a bibliographic reference) in relation to individual transformation processes. It is also important to note that, as in so many other areas of application, technology may well be facilitating the provision of services for which there is no non-technology parallels. Case study 1 explores this issue in more detail.

Technology markets

In general terms, LIS departments are 'at the mercy' of ICT markets and their dominant suppliers, with only minimal in-house development capacity. This has been especially evident in the fields of both library automation and management systems. In-house systems have been abandoned in favour of turnkey ones that have often been less easily customisable than the 'home-made' product. This change has not been seen as having an impact on the competitive edge of individual institutions, although pressures to lever money or recruit users has meant that there is a greater drive to enhance facilities offered. In this context, organisations that engage in 'vertical' relationships with suppliers and collaborations with suppliers are better placed to offer differentiation to their users/customers (e.g. electronic publishing, including as a base for distance learning) than others. At the same time, strategic 'horizontal' relationships (e.g. with other LIS departments in the same region, or covering the same subject areas) allow institutions to harness technology collectively, whether for joint development of content, software, infrastructure or a critical mass of purchasing power with suppliers. This development of a critical mass is particularly valuable where emerging technologies require a standards base. Managed learning environment (MLE) development is a good current example where this is the case.

Cumulative innovation

The move away from in-house development of systems (especially housekeeping ones such as finance, library, personnel records, timetabling) has meant a lack of cumulation of innovation in central ICT management and provision. Until recently, only 'at the edges' (e.g. electronic publishing or teaching and learning support) has there been such a cumulation, and even then, there has been relatively little take up of the technology until now. Institutional politics often means that collaboration within the organisation is limited. LIS units, for example, may well have a better level of successful collaboration with local and national organisations than with other parts of their parent body. Where LIS departments are involved with R&D projects with external partners over a period of time, there has been a cumulative development of expertise in, for example, electronic document delivery, although it has taken a considerable time for these new products to be fully integrated into mainstream activity. Where this is happening, it is a result of further development by commercial partners rather than in-house work by LIS departments for the simple reason that they (or their parent bodies) do not have the capacity to maintain and develop further the systems past the prototype stage.

Acceptance and intervention

Senker (1990) talks of two kinds of technology departments: acceptors and interveners, the first being reactive and the second proactive. Accepters 'carry out the minimum of tasks necessary to cover themselves under ... [the] law'. Interveners 'cover a much wider range of activities ... [providing] in-house services ... [imposing] requirements and ... [diversifying] in new directions'. This proactivity on the part of interveners includes links with research organisations and partnerships with suppliers, although the latter two groups tend to be the more dominant partners.

In broad terms, LIS departments are typically accepters rather than interveners. Legislative requirements are not a strong driving force to intervene, except perhaps in the area of health and safety at work compliance, where legal minima have to be observed. LIS departments are heavily reliant on external suppliers, with in-house technology departments (whether within the LIS area or the parent body, at least in public sector organisations) having been cut back or closed as a result of financial stringency. The purchasing power of individual LIS departments is too small to carry much weight with technology suppliers, although strategic alliances with other LIS organisations does normally mean that combined purchasing power enables procurement officers to have some technical input into suppliers' developments of technology supplied to LIS units.

In the UK and some other western countries, there are areas where LIS departments have been interveners rather more than accepters. One obvious example is in the field of electronic publishing. It has often been LIS staff who have taken the initiative in terms of persuading publishers to move away from print-on-paper to electronic publication, with several alliances with suppliers (publishers), where the LIS departments (and their parent bodies in the case of HE institutions) have either strongly influenced the publisher label or, in the case of courseware, virtual learning environments (VLE) and MLE development, have begun to develop their own labels. This intervener status has also led to alliances between groups of institutions to form a critical mass of interest in, and knowledge of, ICT. This enables the customers to deal with the suppliers on a more equal footing than would otherwise be the case.

Supplier power

Until recently, individual library users have not been powerful 'buyers' because there was little competition within the LIS sector. However, the emergence of Internet-based information suppliers has provided a genuine alternative to the traditional LIS model. On the other hand, the power of libraries as buyers has been growing stronger in many areas – whether data or materials – not least because of the ability of cooperatives to buy in bulk, underpinned by a drive to reduce purchasing costs born of the non-profit-making background of most LIS units, coupled with a significant decrease in the unit of resource available to them in recent years.

The power of suppliers in the LIS sector has been particularly strong in those environments where a dominant supplier or suppliers has 'cornered the market'. In the UK, for example, the British Library has long been the major force in document delivery. It had until recently a unique service that represented good value to organisations keen to reduce their own high fixed costs (as for example with storage space) and from which it would have been expensive to decouple. However, the increased availability of alternative sources of supply, not least through more advanced and differentiated use of technology, has increased the power of the buyer (whether individual user or LIS unit) and reduced the dominance of the key supplier.

The case of the British Library is a good example of the potential threat of substitute products or services, in this case through the innovative application of technology to LIS activities, eventually breaking the traditional dominant design. This issue is explored more fully in Case study 1. It does highlight the need for organisations to manage technology strategically and in the context of the overall environment in which they are operating.

Endnote

This, then, is the background to the following case studies, which relate to various aspects of strategic technology management. They do not necessarily represent good practice. Instead, it is suggested that they are best viewed as 'highlight' or 'lessons learned' reports against which the key recommendations and proposed best practices from the book can be tested.

References

Feather, J. (2003) 'Theoretical perspectives on the information society', in S. Hornby and Z. Clarke (eds), *Challenge and Change in the Information Society*. London: Facet, 3–17.

Hornby S. and Clarke, Z. (2003) *Challenge and Change in the Information Society*. London: Facet.

Murdick, R.G. et al. (1990) *Service Operations Management*. Boston: Allyn & Bacon.

Robinson, B.M. and Robinson, S. (1994) 'Strategic planning and program budgeting for libraries', *Library Trends,* 42(3), 420–7.

Senker, J. (1990) *A Taster for Innovation: British Supermarkets' Influence on Food Manufacturers*. Bradford: Horton Publishing.

Snyder, H. and Davenport, E. (1997) *Costing and Pricing in the Digital Age: A Practical Guide for Information Services*. London: Library Association Publishing.

Case study 1

ICT and document delivery in the UK

Introduction

This case study looks at the application of information and communications technology (ICT) within library and information services (LIS) in UK higher education (UKHE). It concentrates on assessing the ways in which document delivery has been transformed into a central driver in Internet-based library service developments and considers recent innovations and likely future pathways. It focuses on the results of programmes such as the Joint Information Systems Committee (JISC)'s e-Lib Programme and subsequent attempts to provide radical alternatives to the dominant design of traditional inter-library loan and document supply services. It draws on first-hand experience of the innovation process through projects such as EDDIS and Agora.

I argue that the old dominant design of document delivery, ably supported by national/centralised organisations such as the British Library, was breaking down, at least in terms of journal article supply. 'The more things change, the more they can never be the same' (Veaner, 1982). But it is not just a question of major change within the 'transformation process' that turns a document request into a document delivered. I also argue that the likely future model for document delivery is emerging in a significantly different *context* from the old-style one based on inter-library lending (ILL). And it is a context not just of pervasive connectivity to and use of the Internet but also, partly as a result of web developments, of document delivery as *the* central element in support of research, teaching and information access, and knowledge management more generally. In addition, all this change is taking place – in the UK at least – in an environment where the whole tertiary education system has changed – and will continue to change – radically. It is certainly a fluid time in terms of document delivery issues. As JISC has already recognised, 'we are entering a very important new phase of development and need to acknowledge both how far we have come with the development of electronic libraries and learning resources, and also how far we have to go' (JISC, 2002).

Background

UKHE is a complex and varied sector (Dearing, 1997). Yet there is a pervasive common denominator: the increasing use of ICT to support teaching, research, managerial activities and the delivery of academic support services such as the university library (JISC, 1995; HEFCE, 1997). UKHE has had to respond to the challenge of continuous IT development (Dearing, 1997 – see especially recommendation 42). University staff must change working practices, develop new skills and harness ICT to best effect. This will require much more flexible project working (Anstey, 2000).

Within UKHE LIS, all major housekeeping activities have long been automated (Rowley, 1993). More recently, the Internet has revolutionised the ways in which libraries are perceived. Physical barriers have been removed and the resource base has been broadened out to global levels (Follett, 1993). Information is now routinely retrieved electronically and primary source materials are increasingly stored and accessed digitally rather than in hard-copy format (JISC, 2001). Table CS1.1 gives a rough chronology of the various stages of ICT application in UKHE LIS.

Table CS1.1 A chronology of ICT in LIS

Phase	Description	Comment
Late 1960s	Initial R&D work	Concentrated on circulation and bibliographic services; aimed at improving efficiency of existing processes and services.
1970s	Productisation	Introduced circulation and bibliographic services to a wide number of libraries; development of a range of library suppliers competing within the LIS market.
Early 1980s	Extension of products	Extension to acquisitions and document delivery to ensure products remained competitive.
Late 1980s to mid 1990s	End-user developments	A new departure; increasing emphasis on end-users. The beginning of networked catalogues, personalised access to bibliographic/personal information, increased self-service. Changes were intended to enhance products, improve value for money and increase efficiency. Library housekeeping systems largely mature, through a new subphase with the development of graphical/Web interfaces in response to rapid and widespread use of the World Wide Web.
Mid 1990s	The Internet revolution	A move away from housekeeping systems and public access to them towards a new construct of LIS provision based on remote, digital and virtual provision. The sector has moved back towards invention and innovation rather than continuous improvement and reinnovation.
2000s	New constructs	A deconstruction of the library as we know it; the development of new constructs of library and information provision, fully integrated with other aspects of learning, teaching, corporate information and individual information management that is completely separated from physical space requirements; in retrospect, the period will be seen as one of 'discontinuous change'.

Changing the dominant design

There has long been a 'dominant design' in LIS – that of hard-copy storage, access and delivery – despite (or perhaps because of) the many process innovations over the years. However, by the time the e-Lib programme started in 1994, the UKHE LIS sector had reached that final 'specific pattern stage of innovation ... [where it was] vulnerable to the possibility of a revolutionary new product introduction' (Noori, 1990), as noted in Table CS1.1. In this case the digital library threatened to overturn the old dominant design for both library provision and document delivery. The difference between interlending/document supply and on-site provision disappears when access to any resource is virtual rather than physical and when distance or proximity to the end-user is of no consequence; in this emerging environment, the role of the librarian as gatekeeper and intermediary was set to evaporate.

But the dominant design was not just breaking down in terms of the process of document delivery; what would a document look like? In a 1998 article (Baker, 1998) I wrote:

> What is an electronic document? It is here defined as information presented electronically with the same objective as a paper equivalent and additionally, information stored electronically in multimedia form which may or may not include graphics (static), video (moving images), sound, text as hypertext (e.g. SGML or even ASCII), software, datasets, etc.

We talk increasingly not of 'documents' but of 'information assets'. There is much added value from digital storage and delivery that this not obtained from the old-style print-on-paper format. A further quotation from my 1998 article:

> Why is [the World Wide Web] so popular? It has a number of attractive characteristics. It is: multimedia; interactive;

> accessible; user-friendly. 'It has a facility for displaying objects of several non-textual types, and for preserving the unique "look-and-feel" of objects emanating from different sources'. The ability to embed hypertext links within the body of 'information objects' makes it highly interactive – a feature not offered by other retrieval systems. Its accessibility is almost endless; the Web is a global information system. It 'permits the unrestricted dissemination and use of sets of information objects of all types'.[1]

Arguably the most important reason why the Web is proving to be so popular is its user-friendliness. The end-user does not have to understand the various protocols that are used in the access and transfer of the different types of object available over the Internet. The Hypertext Transfer Protocol allows for the seamless retrieval of a wide range of electronic documents, from anywhere in the world, at the click of a mouse.

It is argued, then, that within the UKHE environment, the development of Internet-based resources and services and the breakdown of the old dominant design is a 'landmark' change (Kingston, 2000). In Table CS1.1 above I have described it as a 'revolution', a case of 'the more things change, the more they can never be the same' (Veaner, 1982). On the one hand, new technology is being applied to existing markets; on the other, new markets may be tapped once the products and services that are emerging from the present revolution are fully developed. Existing suppliers, such as traditional libraries and document delivery services (and, indeed, users) may find it 'difficult to adapt to environmental changes with ... an ageing product' (Noori, 1990). 'For the first time in history, children are more comfortable, knowledgeable, and literate than their parents about an innovation central to society' (Tapscott, 1998). This is certainly causing problems of staff management, training and development in UK universities (Anstey, 2000).

The ICT 'revolution' is having a particular impact on the traditional inter-library loan service within UKHE libraries. This area of activity is gradually being transformed from a specialist, stand-alone service to one that is all pervasive and centrally integrated into mainstream library activity to the point where it is the central driver of strategy, policy and practice. Table CS1.2 suggests some of the ways in which the transition from old to new styles of document delivery is happening.

Table CS1.2 Old vs. new style document delivery

Old-style inter-library loans	New-style inter-library loans
Largely print on paper	Largely digital storage and delivery
Minimal integration with other library operations (e.g. separate user database)	Maximum integration with other library operations
Restricted to a small proportion of the campus population	Available to, and used by, the whole of the campus population
Provided primarily to support research	Provided to support learning and teaching as much as research
Intermediary (i.e. ILL librarian) driven	End-user driven
Secondary means of support	Primary means of support

As the original Agora project description (*http://hosted.ukoln.ac.uk/agora/*) put it:

> Libraries are exploring this transition, but it will take some time to evolve routine services and practices. This is because existing services and practices are oriented around the library as place. However, more importantly, it is because the environment is uncertain, information providers are undergoing similar transitions, and a mature technical and commercial infrastructure for networked information does not yet exist.

Reverse product cycles

Much of the more recent development noted above has been brought about within the UK at least in part by the significant programmes of work commissioned and overseen by the JISC, though developments in North America and Australia have also been substantial (see, for example, Greenaway, 1997). The antecedents of the e-Lib programme were those continuous improvement processes that started with library housekeeping systems in the late 1960s, as noted above. A 'reverse product cycle' thus started, with the introduction of 'the simplest and most incremental process innovations ... aimed at improving the efficiency and reducing the costs of delivery of existing products' (Barras, 1990). By the early 1990s, this reverse product cycle was affecting inter-library loan or, more particularly, journal article supply and a number of experiments were being carried out, initially with the aim of speeding up the delivery process (Baker, 1994a, 1994b).

Even then, however, there were the beginnings of a new, more integrated, more innovative approach to document delivery that sought to bring all the various elements together under the control of the end-user rather than the inter-library loans librarian (Baker, 1991, 1992a). Libraries not only needed

> to work smarter and more efficiently for the benefit of an identified clientele, but also [needed] to be innovative in the way services are provided to make the best possible use of resources: human, material, technical and financial. Traditional approaches to providing interlending and document delivery services have typically resulted in operations with little or no integration of systems and with little opportunity to improve efficiencies at the point where the services cost most – library staff. (Greenaway, 1997)

Because the reverse product cycle was gathering radical momentum, and because the traditional paradigm of research library support in particular was breaking down under the weight of journal inflation and government funding cuts (Baker, 1992b), the e-Lib programme had fundamental innovation at its core. R&D programmes like e-Lib typically seek to stimulate 'process discontinuity' in order to create 'a very different and much more competitive institutional régime' (Barras, 1990; see also Follett, 1993) than was hitherto the case, even though there may be casualties along the way. These casualties may be existing suppliers or ways of working, as already noted, or may relate to the R&D effort itself. A standard paragraph in JISC calling for proposals for projects, for example, states that there should be a 'recognition that in groundbreaking work there may be failures as well as successes, but that all such experience can provide valuable information for the [UKHE] community' (JISC, 2002). And some of the e-Lib programme projects failed, at least in their initial objectives, even though they allowed the sector to learn many valuable lessons (EDDIS Consortium, 2000). But whatever the outcomes, there was a sense of no turning back: the old library paradigm would simply not work any longer.

New visions, new concepts

So new visions began to emerge. The CILLA project, for example, envisaged 'system development ... which automates as much as possible of the internal procedures relating to interlending and document delivery, while facilitating the move from wholly mediated services to a point where library staff need only intervene under specified conditions, usually to tackle exceptions.' In order for this to happen 'innovative reorganisation or reengineering of those processes' would be required (Greenaway, 1997). Cooperative alternatives to the British Library's centralised document supply service model such as COPAC and LAMDA emerged.

Some of these cooperatives took the form of 'clumps'. The clump concept was a response to increasing problems of service discovery and navigation as more and more networked services came on line. Clumps of data services that shared common features would be organised around a specific geographical location, a subject specialisation or an intellectual domain. Clumps could therefore help to organise the content of the network and help searchers to select the right databases for particular searches. Clumps might also represent networked services in their own right; for instance, in the library community, a clump of databases might be formed to provide a logical union catalogue, or to support a regional inter-library loan service.

At the same time, commercial organisations were also developing new models. In 1997, for example, publicity from SilverPlatter described the company's full content strategy, 'SilverLinker'. This enabled researchers to navigate seamlessly to full content. 'SilverLinker offers a superior approach based on customer access to hundreds of databases and potentially unlimited numbers of journals. This powerful service is completely natural for end-users – identifying a specific article when searching across numerous bibliographic databases, then requesting the full article itself.'

More recently, a joint working party of the Publishers Association and JISC recommended that a pilot project be funded 'to provide an alternative to inter-library loan for the supply of electronic copies of journal articles, whether these are retained in electronic or only in print form. This alternative route ... would be from the publisher of the journal article via a clearing-house service at a standard price no more than libraries currently pay for an inter-library loan request.' In a pilot project a number of publishers and libraries would establish procedures, test systems and evaluate results. Electronic requests would be received from libraries. There would then be a process to establish whether or not electronic delivery is available (and, if not, redirect the request to the normal paper-to-paper system). The source of the article would be identified; the article requested from the publisher or other supplier. The text would then be transmitted to the user (via the library if necessary) with the library being notified of delivery. The library would be invoiced for articles supplied and an agreed proportion of the income distributed to the publisher (*http://www.jisc.ac.uk/pub99/jp-edd-prop.html*).

The e-Lib programme itself had a major 'action line' concerned with document delivery that was to fund Infobike, EDDIS and SEREN – three major projects that were expected to make a significant difference to traditional inter-library loan (further details can be found at *http://www.jisc.ac.uk/elib/projects.html*). Within this line, the EDDIS (Electronic Document Delivery – the Integrated Solution) project provided a particularly full description of the likely new dominant design. EDDIS was a major three-year project aimed at developing a seamless electronic bibliographic search, item locate, order, document delivery and account service for teachers and students in higher education.[2]

Integrative approaches

The underpinning theme of these developments was of technology and process integration. The potential for linking OPAC data with other sources of information and documents was recognised as early as the mid-1990s (Leeves, 1995). The EDDIS Annual Report for 1997/8 (EDDIS Consortium, 1998), for example, asserted that: 'Information access in universities is a series of independent processes, often mediated by the library. EDDIS enables integration into a single process, which is user-driven but library-controlled. The process is complex and variable between institutions providing much scope for simplification and resultant benefit.'

The development of a new construct of document delivery in the new information environment could be stripped down to the age-old basic input–transformation–output process (Figure CS1.1). Although this was as valid for a chained library as a virtual one, it was the transformation process that required a radical overhaul. The EDDIS project, for example, concentrated on a number of secondary input/output measures to produce a transformation process that would work in the new environment and the new context that were now being created (Figure CS1.2).

Figure CS1.1 Input–transformation–output process for document delivery

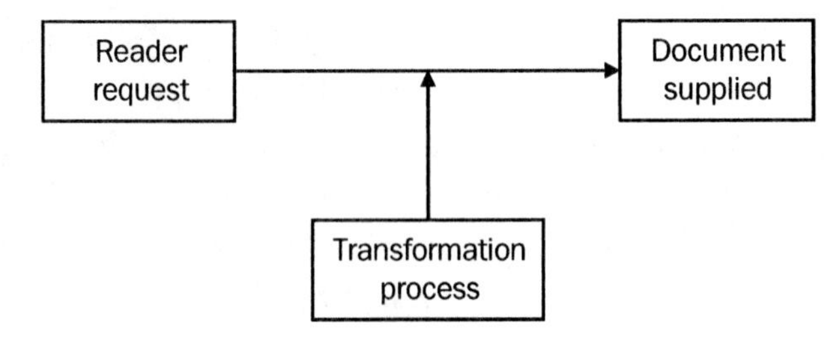

Figure CS1.2 Input/output measures of the EDDIS project

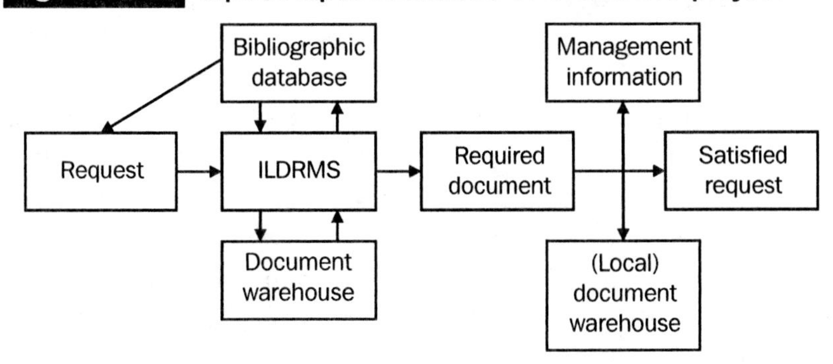

The bibliographic database would tell both the requestor and the library service of the existence and possible location of documents. The document warehouse could be either local or remote; it might consist of several different warehouses. Once the request was satisfied, information would be produced that helped management to monitor and modify services. For example, heavily requested documents initially supplied remotely could then be better supplied locally. Supplied documents could be stored in a local warehouse for general rather than individual usage. The ILDRMS (Inter-Lending and Document Request Management System) represented the transformation technology by which the request would be satisfied. In addition, the supplied documents could be the products of electronic publishing systems whereas the bibliographic data were typically created, manipulated and communicated automatically.

Initially, development work in EDDIS concentrated on alternatives to the traditional model of document delivery – the automation of a process that handled the identification, location, order, receipt and accounts management of books and (photocopies of) journal articles. The project team was nevertheless aware of the need to plan longer term for both scan-and-send requirements and full electronic publications.

The innovation process

The kind of innovation that e-Lib and EDDIS-type programmes and projects have sought to stimulate has a number of aspects. Above all, it has invention at its root. In the case of JISC developments over the last five years, the inventions have been driven particularly by concepts such as the electronic, digital or hybrid library, including the development of protocols and standards for digital library application. This has all been within the context of joined-up institutional strategies and distributed but coherent and complementary national information environments (*http://www.jisc.ac.uk/pub99/dner_vision.html*).

Inventions require research and development work to create them and then prototyping to provide practical demonstrations of the concept. At the prototyping stage there is typically still a good deal of work – and investment – to be undertaken, and e-Lib was as much about proof of concept as about working services, whether or not that was the original intention. But the invention/innovation process is almost always very expensive (Kessler, 2000), though to JISC's great credit the e-Lib programme was allocated significant resources (by UKHE standards) for product development. The transformation from 'proof-of-concept' to fully working, marketable, sellable and successful product is even more fraught than the initial development work, not least thanks to issues of scale and critical mass and the need for risk-taking and market strategies. And at this stage, organisations like JISC typically hand over to commercial players to finish the job.

However, without a willingness to take risks, based on the likelihood of sufficient business accruing to the new product to justify the return on investment, new products will not emerge into the market – in this case UKHE LIS. And there remains a strong allegiance to the prevalent 'old-style' dominant design of document delivery in the UK, despite the threats posed by the Internet revolution. The EDDIS project, for example, found the transition from prototype to working product and service difficult if not impossible for these very reasons (EDDIS Consortium, 2000).

The new vision

In 1999, the University of East Anglia (UEA) carried out an extensive Delphi study as part of the rewriting of its information strategy and the development of a programme of work encompassing its information services and the longer-term use of ICT. The study was based on a three-month iteration of comments made by a range of experts in the field of LIS and beyond. Although the study focused on developments within the context of UKHE, the summative statements concerning the future of library services were deemed to be of general relevance:

> *(Respondents were asked about how library and learning resource provision will look in 5–10 years' time and where it will differ most and where it will differ least from traditional book-based provision.)*
>
> One of the difficulties here is that many elements of traditional provision will remain alongside newer approaches. ('For most academics in the humanities, the more it stays as it is, the better'; 'Library/learning resources will be predominantly electronic and predominantly sourced off-campus under access agreements.') With development in e-journals, etc. researchers may become more self-sufficient and visit the library less often, but students are likely to continue to see it as a place to work and get help. Integrating print and electronic provision successfully will require careful planning, and skilled training/coaching of students to ensure that they make full use of both new and traditional media. The range of formats, different options for the same information, web-based delivery, and self-service transactions will be a key feature of the future environment.
>
> Some collection policies may become obsolete. Where they continue, they will focus particularly on: what is paid for up-front and what is pay-per-use; criteria for retention and disposal; what is managed by individuals, groups, departments and 'the centre'. In the longer term, librarians will become organizers of knowledge and information access (particularly

> informally-generated information) and the gatekeepers to information and budgets. Users will interact through this empowerment directly with information owners and not through librarians. Budgets will increasingly devolve to departments and users. In ten years, few people will borrow a real book, but will access its contents electronically, most commonly without visiting the library. This suggests a need for a 'Help desk' function in the library. This will not apply to subjects where the study of a physical document is needed from time to time, like classics, literature, medieval or ancient history or literature. Book swaps are already becoming automated and will be electronic. (Baker, 1999)

In other words, traditional libraries will continue to exist, at least for specialist requirements, alongside digital ones. This long-term 'hybrid future' has been more recently summed up by JISC:

> One of the key lessons learned by the JISC through investment in a range of programmes to enhance access to research and learning resources ... is that we are operating in an environment where users interact with a variety of digital and non-digital objects. In searching for high quality ... resources in networked environments they will encounter an array of electronic records pointing to both original items, a book, manuscript or painting, and records leading to digital and increasingly ... original digital resources, a satellite data stream, multimedia essay, or digital art work. In all of these cases [users] will need to be confident about the quality of the information that they are accessing and have the skills of judgement and understanding in place to assess and utilise what they find ... (JISC, 2002)

But there is an imperative:

> Access to high quality on-line information and learning resources is now essential to all engaged in education, whether as students, teachers, or researchers ... The development of a coherent information environment is an important means of helping users to maximise the value of the Internet, by making best use of its bewildering profusion of information resources. (JISC, 2002)

The new paradigm

If both digital and non-digital resources have their place in 'enhancing the quality of the learning and research experience' (JISC, 2002), then, by implication, we are looking at the further development of a hybrid document delivery model for the foreseeable future. Describing the likely dominant design is not so simple, however.

In simple terms, the hard-copy-based 'dominant design' has broken down for journal articles delivery. It has not done so for books, where there is still a much greater emphasis on print-on-paper, especially for research and more advanced teaching. However, the situation is far more complex than that. There is now a whole spectrum of possible models – and hence no single dominant design – ranging from the most traditional inter-library loan to the most advanced electronic document delivery model yet imaginable. The point along the document delivery spectrum at which a particular reader/user request is satisfied will depend on a whole range of variables, as shown in Figure CS1.3. I will take each of these variables in turn and discuss their impact on the document delivery spectrum.

Figure CS1.3 Variables along the document delivery spectrum

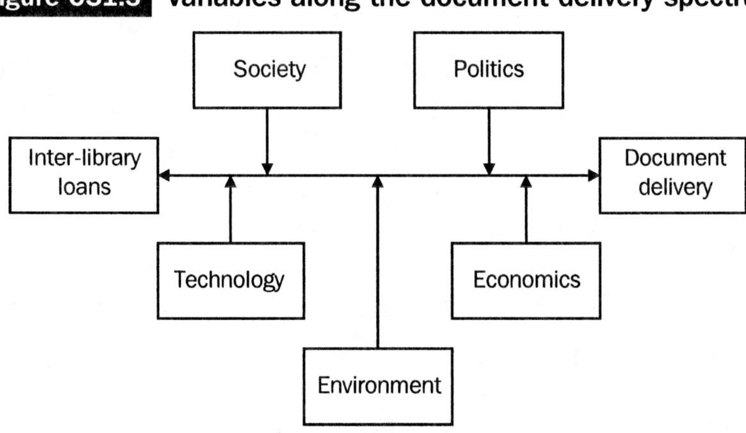

Technology

Given what has been written above and in Chapter 6, technology developments are crucial to any positioning on the spectrum. Internet connectivity is pervasive: 80 per cent of students arrive at the University of East Anglia with a mobile phone and 60 per cent with a laptop. By the time they graduate, ownership of both pieces of kit is almost universal.[3] The pace is already fast, but it is quickening. How many had mobile phones or computers even five years ago? And this continuing emergence of 'new' technologies has the potential to affect all the other variables listed and to move yet more activity to the right-hand side of the model. Digital technology, including its associated communication networks, already has the potential to alter radically the way in which research and teaching/learning materials (texts and a whole range of other forms of content such as images) can be accessed in a widely distributed landscape. Virtual and managed learning environments (VLEs and MLEs) are already becoming commonplace – at least conceptually – in UK higher and further education (MLE Steering Group, 2002). And now we are developing VREs (virtual research environments) to 'support the needs of the e-researcher' (Tiedau and Chambers, 2002).

But the future is not just a distributed one. It is also personalised and customised. Digital content is in increasingly diverse formats and users are ever more demanding when it comes to friendly and fit-for-purpose access. Research within the JISC-funded Virtual Norfolk project, for example,[4] has shown that the take-up of digital learning materials is significantly enhanced if individual end-users can manipulate content for their own learning and teaching requirements. This approach relies on the fusion of services capable of bringing together relevant content, arranged according to numerous different objectives or characteristics, and crucially, *as determined by the user*. Technology tools must therefore allow end-users to access a variety of materials and to generate customised learning or research pathways to them. Integrated user profiling – drawing on data from multiple applications and allowing seamless movement between application portals – is seen as being a key element in the digital transformation of teaching practices. For JISC, this means that 'activity will focus on methods to allow members of the community to build the content that they will access, and to share this in meaningful ways with other colleagues and peers. This activity will build a framework for leveraging our mutual community resource' (JISC, 2002).

In order to deliver this vision, however, it is crucial – as in so many areas of technology application – that there is full *integration* of all the various elements. This requires interoperability across a whole range of technologies and applications. JISC's Interoperability Focus[5] sums up the issues relating to interoperation admirably and I do not propose to repeat them here. Suffice it to say that:

> It has become clear that enhanced interoperability for users will not be achieved without agreement on common semantics to support cross searching. As part of developing the

> Information Environment, the JISC will strive for the cross-sectoral adoptions of standard terminologies, for example for subject, audience level, resource type and certification. (JISC, 2002)[6]

The JISC-funded JOIN-UP cluster of projects aims to provide a framework within which the British Library, EDINA, MIMAS and LAMDA (working with Fretwell-Downing Informatics) will contribute 'separate but compatible and inter-operable component parts' to the discover/locate/request/deliver process that is fundamental to all document delivery systems and services.

> The main challenge for JOIN-UP is to realise the full potential of bibliographic services, many of them sponsored by JISC, by connecting users of those services to services on the information objects of interest to them. Thus, users discover references, are provided with information to locate services on the material (typically journal articles), and are also provided with the means to connect automatically, where appropriate, to request and delivery mechanisms.[7]

In the case of JOIN-UP there is a model of unified discovery/location. However, there is no single model for document delivery in general and single article supply in particular. JOIN-UP, for example, does not explicitly include publisher interests. The EASY (Electronic Article SupplY) project, undertaken by Ingenta and Lancaster University Library, was directly concerned with translating the ILL paradigm into an electronic environment. In particular, it attempted to find a solution to the problem of replicating the economics of ILL in the electronic environment. Much of its focus was on persuading publishers to participate. However, during the course of the experiment (now concluded) a particularly significant issue (of much wider significance than just EASY) was the difficulty of matching requests to items automatically where there is no 'negotiation' with the end user. This mirrors the ILL process where a user sends a request – often with inaccurate or incomplete data – to an ILL department, where human mediation identifies the correct descriptors. The current difficulty with having an online negotiating process is that the budgetary control remains – at the present time at least – with the library. Users cannot therefore complete the request if there is a charge for the item being requested.

A number of other products are now emerging onto the market. One such is Fretwell-Downing Informatics ZPORTAL:

> With a choice of data publishing tools, ZPORTAL offers the freedom to integrate access to the most important and appropriate resources – irrespective of format or location. The new portal solution acts as a one-stop-shop for users, taking them from the initial need for information through to its delivery without having to use several different tools and applications. This means that the frustration that often results from lengthy search processes can be reduced, while efficiency is increased. Where users had many places to look for information they will now have one that can offer them quality local sources plus the best of the web.[8]

Another product is the Ex Libris' SFX Resolution Service, enabling the introduction of linking services. The development of SUNCAT adds a further dimension to this area. SUNCAT will allow the integrated discovery/location of a very large national journal resource and will be a major contributor to the information environment. The same may be true of the CC-Interoperability Project, which will improve access to the resource of the CURL and M25 library consortia. In addition, the January 2003 issue of *Update*[9] contains details of the British Library's 'e-deal' utilising Adobe Content encryption and Adobe Acrobat eBook Reader software to manage the intellectual property rights issue associated with electronic document delivery. Deals have been struck with major publishers like Elsevier and one million PDF files are already available from the British Library.

But it is not just improved access to different types of medium and content that the technologies of recent years have brought. It has changed – or is changing – the whole face of publishing and of the research and teaching/learning endeavour, as for example recent discussions regarding e-prints suggest.[10] These developments have the potential to remove all but author and reader from the publications process or the document delivery loop. There seems little in the way of blockages to the development of a totally digital future if one looks at the technology variable alone. But applying the technology is the easy part. The other five variables have a fundamental impact on the way in which document delivery models will be shaped in the future.

Society

Technology trends and developments have to be seen in the context of a much broader set of societal changes that affect the ways in which new models are adopted. In terms of higher education, many of these broader issues are encapsulated in a recent government White Paper (Department for Education and Skills, 2003). This document contains a whole set of drivers that will affect the ways in which digital document delivery will be developed and used over the next 5–10 years. Table CS1.3 summarises both drivers and their likely impact on document delivery.

This brief initial analysis suggests that the technology trends noted earlier have a fertile environment in which to develop, but that the diversification and fragmentation of models is highly likely to occur.

Table CS1.3 Future drivers of change for digital document delivery

Driver	Possible impact
Diversity of mission	A fragmentation of requirements and, hence, delivery models, in terms both of teaching/learning and research and of the methods of providing and accessing learning resources
Tiering of institutions (research, teaching)	Concentration of demand for research literature into a smaller number of institutions, with a greater resource available to provide access to materials to researchers within those institutions. Conversely, less access to such materials from institutions not in the research tier, not least because of their inability to allocate resources to this kind of access. A concomitant increase in the requirement for access to high-quality teaching materials (see also below)
Regionalisation	Increased collaboration between institutions in a given area, with agreed niches for provision (research, teaching, subject areas) leading to greater regional concentration on resource requirements
Widening participation	Greater emphasis on mass access to literature and self-directed rather than tutor-led study as numbers increase without a concomitant increase in resources
More flexible study	An increased emphasis on e-learning and one-stop-shop approaches to higher education that pre-package resources for seamless access
Third-leg funding	Increased transfer of intellectual property rights to business and industry; increased selling of services (including information access) to small- to medium-sized enterprises (SMEs) within a university's local area of influence?
Greater international recruitment	A greater emphasis on flexible and distance learning, supported by the Internet, to allow for increased in-country provision at undergraduate level and UK-based training and education for overseas students being concentrated more on postgraduate study

Politics

Any system as complex and diverse as UKHE inevitably has a political dimension. This is as true of the document delivery environment as of the teaching and research endeavour that shapes and conditions it. In terms of technology application, the key issue must surely be the number of stakeholders who need to be engaged in order to move the innovation process from prototype to full product or working service. Figure CS1.4 shows some of the key stakeholders who have to be taken into account if a new paradigm of document delivery is to be successfully introduced into UKHE.

Figure CS1.4 **Key UKHE stakeholders for document delivery**

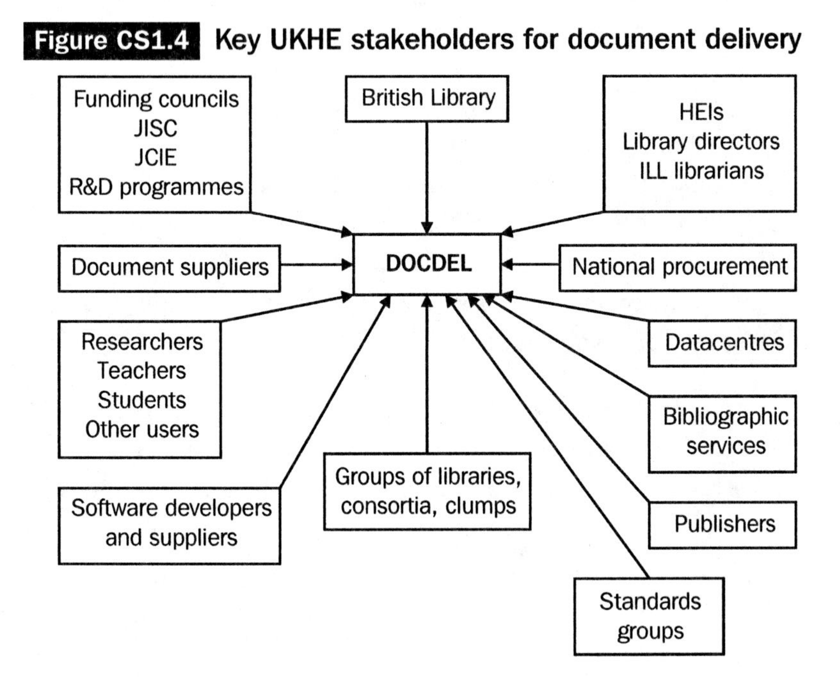

The potential breakdown in the current paradigm of scholarly publishing, especially with regard to journals, and the continuing uncertainty over what the new dominant design will look like economically must make it difficult in some respects for suppliers to engage with the innovation process. And yet:

> ... it is important to bear in mind that it is not only users of electronic services and resources who are going through a process of cultural change in exploring new ways of accessing and using resources. Providers of services of interest to learning and research both commercial and publically funded are clearly also helping to mould and shape this new environment. These stakeholders have a real and vested interest in ensuring that the information and resources they provide can be accessed through and integrated with national frameworks. (JISC, 2002)

Even within the library community, there are political and cultural issues that tend to encourage the setting up of barriers. These militate against the kinds of collaboration and cooperation that are required if end-users really are to have seamless access to what (at the provider end) are a whole series of distributed and disparate sets of resources:

> Growth in genuinely collaborative collection management would probably be the best indicator of deep resource sharing. However, we are doubtful that change can be brought about in this area if any initiative is left as voluntary, and recommend that the RLSG [Resesearch Libraries Support Group] and the funding bodies consider the case for central action and associated incentives. (Schofield et al., 2001)

Economics

Reference has already been made to the EASY project and its emphasis on translating the economics of ILL to eDocdel. However, we remain short of economic models that will help us to plan strategically the management of the document delivery spectrum described earlier. White and Davies (2001) comment that:

> The advantages of electronic access and/or the opportunities for extending access to a wider range of titles is gained at significant cost either in cash terms or as a proportional increase in spend. We all know too well that library budgets are not infinitely elastic and priorities regarding materials and other inputs have to be established. Whether such strategies are desirable, or can be afforded still remain questions to be addressed by individual institutions ...
>
> The entire issue of information and document access will continue to exercise managers on a range of issues – strategic, financial, technical, and operational. The opportunities and choices grow as do the challenges and difficulties of decision-making.

This position is reinforced by Schofield et al. (2001):

> Under current arrangements it is generally impossible for libraries to share electronic resources on any other basis than by aggregating the existing spend, so there are no economies of scale to be enjoyed. There are potential savings to be had from institutions collaborating to form a bigger virtual customer, but there is also a cost in terms of staff time required to manage the process. An objective evaluation of the costs and benefits of different forms of procurement arrangement would be helpful.

Environment

The management of organisational change is arguably the most challenging aspect of the digital document delivery future. There is much evidence that the technology is still outstripping the users' ability to maximise it:

> We can create new environments offering richer access and more meaningful and relevant resources, but how can we stimulate and engender engagement with them? This is not just an issue for the JISC but for all members of the higher and further education community (and beyond) who aim to maximize the benefits of access to digital resources. (JISC, 2002)

Much still needs to be achieved in terms of culture change within higher education. Projects that aim to embed new technology across whole institutions and sectors always carry a high risk of failure where appropriate organisational and cultural foundations have not been well laid. In particular, UK public sector projects have suffered from a mismatch between product conception and environmental reality. Those who were meant to use and benefit from the new systems were simply not ready for or even informed about them (Heeks et al., 1999). 'Attitudes and behaviours among professional librarians, academics and students continue in many instances to be a barrier to the effective embedding of [digital] products and services' (ESYS Consulting, 2001). More recently, the INSPIRAL bibliography (*http://inspiral.cdlr.strath.ac.uk/resources/bibliography.html*) and the ANGEL project's initial formative evaluation (*http://www.ariadne.ac.uk/issue30/angel/*) provide a clear insight into the issues surrounding the implementation of digital learning environments. Transatlantic delivery brings its own significant change management issues, as discussed in the recent GATS report (Knight, 2002).

Summary

This case study has looked at the key variables that will affect the further development of document delivery in a digital environment. Although it has concentrated on the technology applications, it has done so in a broader societal, economic, environmental and political context. And it has suggested that there will be no single future model of document delivery. There will rather be a series of options, driven not only by the variables described here but also by the individual's personal preferences – limited in UKHE terms only by the imperatives and frameworks laid down by the government of the day. As stated in the previous chapter, it will be a challenging, and indeed an uncomfortable time for librarians, libraries, publishers and technologists alike.

Notes

1. Quotations taken from *Vine* 99 (June 1995), 3.
2. The project partners were: the Universities of Bath, Lancaster and Stirling and the University of East Anglia (UEA), and the Bath Information and Data Service (BIDS). Fretwell-Downing Informatics joined the project at a later stage. The British Library agreed to be an associate partner and to supply documents. UEA was the lead partner.
3. Data taken from internal studies of student use of IT at the University of East Anglia over a period of three years.
4. *http://virtualnorfolk.uea.ac.uk/*
5. *http://www.ukoln.ac.uk/interop-focus/*
6. See also *http://www.jisc.ac.uk/dner/development/guidance/DNERStandards.html*
7. Text taken from the original JOIN-UP bid to JISC.

8. Background on the Scholar's Portal Project is available at: *http://www.arl.org/access/scholarsportal*
9. *Update* (2003), 2(1), 8.
10. *http://amsci-forum.amsci.org/archives/september98-forum.html*

Bibliography

Anstey, P. (2000) *C & IT Skills: Developing Staff C & IT Capability in Higher Education.* Norwich: University of East Anglia.

Baker, D.M. (1991) 'From inter-library loan to document delivery', *Assignation: ASLIB Social Sciences Information Group Newsletter*, 8, 24–6.

Baker, D.M. (1992a) 'Access versus holdings policy with special reference to the University of East Anglia', *Interlending and Document Supply*, 20(4), 131–7.

Baker, D.M. (1992b) 'Resource allocation in university libraries', *Journal of Documentation*, 48, 1–19.

Baker, D.M. (1994a) 'Document delivery: the UEA experience', *Computers in Libraries International 1994: Proceedings of the 8th Annual Conference.* London: Meckler.

Baker, D.M. (1994b) 'Document delivery: the UEA experience', *Vine*, 95: 12–15.

Baker, D.M. (1998) 'The multimedia librarian in the twenty-first century: The viewpoint of a university librarian', *Librarian Career Development*, 6(10), 3–10.

Baker, D.M. (1999) 'The Future of Information Services at the University of East Anglia: The Final Report of the Delphi Project.' Unpublished paper.

Barras, R. (1990) 'Interactive innovation in financial and business services', *Research Policy*, 19, 215–37.

Dearing, R. (1997) *Report of the National Committee of Enquiry into Higher Education*. London: Funding Councils.

Department for Education and Skills (2003) *The Future of Higher Education*. London: DfES.

EDDIS Consortium (1998) *Annual Report*. Bath, Lancaster, Norwich and Sheffield: EDDIS Consortium.

EDDIS Consortium (2000) *EDDIS Extension Project: Final Report*. Bath, Lancaster, Norwich and Sheffield: EDDIS Consortium [the report includes the final report of the original EDDIS project].

ESYS Consulting (2001) *Summative Evaluation of Phase 3 of the Elib Initiative: Final Report Summary*. London: ESYS Consulting.

Follett, B. (1993) *Joint Funding Councils' Libraries Review Group Report*. London: Funding Councils.

Greenaway, J. (1997) *The Coordinated Interlibrary Loan Administration Project [CILLA]: Final Report & Recommendations of the Feasibility Study*. Canberra: AVCC.

Heeks, R., Mundy, D. and Salazar, A. (1999) *Why Health Care Information Systems Succeed or Fail*. Manchester: University of Manchester Institute for Development Policy and Management.

HEFCE (Higher Education Funding Council For England) (1997) *Information Technology Assisted Teaching and Learning in Higher Education*. London: HEFCE.

JISC (Joint Information Systems Committee) (1995) *Guidelines on the Production of an Information Strategy*. Bristol: JISC.

JISC (Joint Information Systems Committee) (2001) *The Distributed National Electronic Resource*. Bristol: JISC.

JISC (Joint Information Systems Committee) (2002a) *Circular 1/02: Focus on Access to Institutional Resources Programme*. Bristol: JISC.

JISC (Joint Information Systems Committee) (2002b) *Information Environment: Development Strategy, 2001–2005 (Draft)*. London: JISC (*http://www.jisc.ac.uk/dner/development/iestrategy.html*).

Kessler, E.H. (2000) 'Tightening the belt: methods for reducing development costs associated with new product development', *Journal of Engineering and Technology Management*, 17, 59–92.

Kingston, W. (2000) 'Antibiotics, invention and innovation', *Research Policy*, 29, 679–710.

Knight, J. (2002) *Trade in Higher Education Services: The Implications of GATS*. London: Observatory on Borderless Education.

Leeves, J. (1995) *Library Management Systems: Current Market and Future Prospects: A Report Prepared for the SCONUL Advisory Committee on Information Systems*. London: SCONUL.

MLE Steering Group (The Managed Learning Environments' Steering Group of the JISC) (2002) *A Report on the Interoperability Pilot Programme*. London: MLE Steering Group.

Noori, H. (1990) *Managing the Dynamics of New Technology: Issues in Manufacturing Management*. Englewood Cliffs, NJ: Prentice Hall.

Rowley, J. (1993) *Computers for Libraries*. London: Library Association Publishing.

Schofield, A. et al. (2001) 'Barriers to resource sharing among higher education libraries: a report to the Research Support Libraries Programme (RSLP)', *New Review of Academic Librarianship*, 7, 101–210.

Tapscott, D. (1998) *Growing Up Digital: The Rise of the Net Generation*. New York: McGraw-Hill.

Tiedau, U. and Chambers, S. (2002) 'Developing the virtual research environment', *Relay*, 54, 22–4.

Veaner, A.B. (1982) 'Continuity or discontinuity – a persistent personnel issue in academic librarianship', *Advances in Library Administration and Organisation*, 1, 1–20.

White, S. and Davies, J.E. (2001) *Economic Evaluation Model of National Electronic Site Licence Initiative (NESLI) Deals*, LISU Occasional Paper 28. Loughborough: Library and Information Statistics Unit.

Case study 2

ICT take-up in UK higher education

Introduction

This study concerns the low take-up of Information and communications technology (ICT) in teaching and learning activities within UK higher education (UKHE). The national-level situation is summarised in a Higher Education Funding Council for England (HEFCE) Report (1997).

Background

Although ICT is pervasive in higher education institutions (HEIs), its usage has until recently in many organisations been primarily for administrative or 'housekeeping' activities rather than the delivery of teaching or the support of learning. This is despite the fact that:

- UKHE has long had a high-quality wide area network – the Joint Academic Network (JANET) – and sophisticated local, metropolitan and regional area networks, with widespread campus connectivity in almost all institutions

- There has been a significant investment in the development of ICT-based teaching and learning aids (courseware, computer-assisted or computer-based learning packages, websites, portals and gateways, virtual and managed learning environments, etc.), notably through a series of national initiatives within the HE community, and primarily the Computers in Teaching Initiative (CTI), the Teaching and Learning Technology Programme (TLTP) and a whole series of programmes commissioned and funded by the Joint Information Systems Committee (JISC).

Underlying the core problem of technology take-up in teaching is a series of issues that continue to require a systematic response. These are:

- *The need for competitiveness.* HEIs are increasingly in competition with each other for revenue through students, research and business activity. At a time when students are increasingly looking to part-time study and international students are less able to fund long-term placements abroad, organisations that can offer study in the home using ICT are likely to be able to do so at a competitive rate when compared with the traditional HEI's method of delivery requiring relatively high levels of face-to-face contact and expensive tutorial and study space.
- *Organisational limitations.* Traditional HEIs are consensus-based organisations that pride themselves on collegiality. This has tended to militate against rapid responses to increasing competition (as described above). Additionally, academic management in many of the older HEIs is carried out on a rotational basis; there has often been little incentive to plan strategically in a three-year post, not least when the academic manager remains under pressure to research and publish as an individual. Many universities have a devolved budgetary structure that militates against central initiatives and leadership.

Any innovation is carried out at the level of the individual academic unit rather than the institution as a whole.

- *Ignorance of technology issues.* The Dearing Report (1997) on the future of higher education recognised that there has until recently been a significant knowledge gap in HEIs with regard to ICT, the recommendation at that time being that senior managers needed to be developed who have 'a deep understanding' of ICT issues.

Identifying the issues

This national-level reporting of issues stressed the fact that the issue of low technology take-up was being made apparent by a comparison between certain stakeholders' objectives (e.g. those who fund TLTP-type programmes) and their actual fulfilment.

Follett (1993) and Dearing (1997) both related the need to develop new methods of teaching and learning delivery to the pressures of increased and more demanding markets and reduced financial resources. Their paymaster (the Higher Education Funding Councils) has seen ICT applications as being a way of gaining productivity and maintaining quality at a time when there are fewer members of teaching staff because of economic cutbacks. It is the seeming lack of payback on the significant investment that has been one of the prime ways in which the problem described here has been brought to the fore.

At the same time, however, other stakeholders' interests have affected the take-up of ICT in teaching. The individual organisation is under local pressure to succeed. Many universities are keen to improve their performance in research rather than teaching. Key stakeholders here include the senior management, whose interests are served not so much by quality assessments in teaching (which to date have not significantly increased income) but by research. This places staff (another group of stakeholders) under pressure to increase quantity and quality of research rather than spend time on teaching, and applying ICT to teaching takes time.

As Cooper and Kleinschmidt (1987) comment, there is little attention paid in discussions about topics such as ICT take-up in teaching to the actual customers, in this case the students at the receiving end. However, there are two other groups of stakeholders in the ICT in teaching and learning model. The first is a series of inter-unit subject-based invisible colleges that are a feature of HE activity. They have shown at least some interest in the field in recent years. Additionally, there are the particular enthusiasts for the medium who (regardless of subject), along with the funders, have drawn most attention to the problem of take-up.

While the funders are looking to a return on their investment, both for economic and for political reasons (they cannot afford to be seen to have failed, having spent large sums of money), there is also the question of ethics. Is it right and proper that 'taxpayers' money' should be spent on large-scale projects that do not yield benefit to the community that funded them? In this context, it could be argued that academics' wish to retain quality face-to-face interaction with students is a blockage to serious and widespread ICT take-up, given its oft-perceived self-service, lower quality feel (cf. Hackett, 1994). These issues were further explored by the Higher Education Funding Council for England (1997).

At the institutional level, there is another group of stakeholders that has drawn attention to the problem. These are the ICT specialists whose job it is to implement ICT strategies consonant with the institutional mission. Without a clear steer from the university executive, they are unable to develop the strategy in a meaningful way.

The problems already identified are exacerbated by the short-termism that has been prevalent in UK HEIs as a result of continuing funding cutbacks. Hayes and Abernathy (1994) remark that ‘although innovation, the lifeblood of any vital enterprise, is best encouraged by an environment that does not unduly penalize failure, the predictable result of relying too heavily on short-term financial measures – a sort of managerial remote control – is an environment in which no-one feels he or she can afford a failure or even a momentary dip in the bottom line.’

Applying technology strategies to best effect

There is, then, a need to see how future ICT developments in teaching and learning can take advantage of experience to date and in particular to apply technology strategies to ensure that maximum cost–benefit and greatest levels of user satisfaction are achieved. As Green (1994) comments:

> The structural static perspective divides up the organisation and its external environment into separate bits: culture and organisation lagging behind strategy, lagging behind environmental change. The task of the strategic management process is to overcome this by engineering a better fit between culture, strategy, systems, structure, etc. In this way the organisation can be made to adapt to the exigencies created by an environmental change.

In Earl’s (1989) terms, universities are still in the first phase of the widespread acceptance of technology, at least in terms of teaching and learning, although ICT is pervasive enough to be a source of conflict within institutions. Only in recent years, with senior management recognising the importance of ICT in teaching and learning, is there a discernible move to a second, more delivery-orientated stage, where ICT can be exploited seriously for competitive advantage, with investment in the area being aligned with business strategy in a meaningful way.

Objectives

Universities are by their nature diverse institutions, both within themselves and from each other. They all share, at least in theory, a relatively small number of key objectives, framed at least partly by the dominant stakeholder, the government funding bodies. What has not happened to any great extent, however, is the convergence between the primary objectives (to which most organisations would subscribe) and the strategic and operational objectives of the individual institutions. This is at least partly because, in Schoemaker's (1993) classification, universities are within the political model or, indeed, the contextual view rather than the unitary actor or organisational models. This may not be a barrier in itself, however. The great need is for consensus.

Holistic approaches

It is difficult to conceive of a situation in which the HE community at large could act in a holistic way in terms of ICT applications in teaching and learning, given the differing backgrounds, academic diversity and relative autonomy of individual universities. As Hackett (1994) comments, however, it is important not to graft new technologies on to old (as has happened in ICT-based teaching and learning to some degree) but to look afresh at the issues and the appropriate applications as a whole. The JISC's (1995) *Guidelines on the Production of an Information Strategy* suggested such an approach. The Guidelines have since been acted upon in a large number of HEIs, not least because the dominant stakeholder (in this case the Funding Councils) required that this be the case. What still needs to happen is a real coordination between the collective and the institutional R&D effort in this area and universities' mainstream teaching activities, even though there is also an increasing competitiveness within the sector.

Change management

This gap could be filled by the kind of systematic approach to the management of technical change put forward by Twiss and Goodridge (1989). The authors require that there is a commitment to innovation and that a number of key attributes can be detected within the organisation, including long-term orientation, flexibility to enable a rapid response, creativity and a responsiveness to new ideas. On the face of it, such attributes are all present to some degree in universities. However, it is not clear that an appropriate environment has yet to be created for the effective application of ICT to teaching and learning. This is because the core, interrelating elements required by the Twiss and Goodridge model are not necessarily present to a sufficient degree. In particular, the corporate university culture is bound by the collegiality referred to earlier, whereas the 'attitudes, motivations and contributions of individuals' are just that: they are individuals rather than groups of teachers working towards a common set of university aims. This suggests that the dissemination and institutional ownership of the corporate strategy is inadequate in terms of making radical change 'stick'.

Auditing

Crawford (1991) and Twiss and Goodridge (1989) both propose audits of institutional culture, innovation and technical issues as a means of assessing the necessity for change. In the case of universities, one key problem to be faced when undertaking such audits is the nature of the product portfolio. Universities teach students in order to produce graduates: that has always been a key aim and, one assumes, always will be. In the case of ICT applications, therefore, the important point to bear in mind is the extent to which such applications will process the kinds of graduate that the market requires. In this context, the emphasis will need to be on the inculcation of the required or desired skills and knowledge. Cooper and Kleinschmidt's (1987) model of product outcomes could be useful here. Clearly the strongest success factor of product advantage is as valid in HE as in the commercial world.

Benchmarking

One response is to apply benchmarking to teaching and learning delivery within universities. This has already been happening as part of a wider drive to greater accountability among institutions, in terms of both research assessment exercises (RAEs) and teaching quality audits and assessments. As yet, however, there has not been a systematic review of the application of ICT in teaching and learning, though pressure is now building towards such an approach, not least as a result of the Dearing Report (1997). In this context, it is interesting to note that although the Dearing Committee looked extensively at ICT applications in other countries (notably North America) there appears to have been little effort expended on looking at ICT usage outside HE. However, it is clear that this broad benchmarking has made senior figures within UKHE wake up to the fact that radical change could be on the way and traditional universities need to look closely at how to meet competition from virtual universities.

The wider context: complementarity and contrast

How does this picture of UK HEIs compare with elsewhere? North American equivalents would seem to be ahead in terms of the application of ICT to teaching and learning, whereas European counterparts, in general, lag behind even the UK. There are several reasons for this. If one takes Twiss and Goodridge's (1989) technical change analysis template as a basis for comparison, it is possible to discern a number of differences between the UK and North America/Europe/the rest of the world.

Cultural differences must play a part, both at national and at organisational levels. In many European countries, for example, the emphasis on lecture rather than classroom format in universities coupled with the traditions of Marxist–Leninist approaches to higher education in the former Eastern bloc have resulted in differing attitudes towards new approaches to teaching and learning (e.g. people still expect to be taught rather than to learn, whereas ICT-based learning tends to emphasise learning rather than teaching). At the same time, such countries have been eager to catch up in terms of education. The same has been true of South-East Asia, although here there is arguably a greater degree of commitment to radical change and a more centralist direction, as well as perhaps a higher dedication to learning.

Innovational record at the macro level inevitably impacts on the technology take-up within universities. Some countries have an enviable record in terms of innovation, and this must influence university education. In South Korea, for example, there has been the development of a 'smart' school, college and university system. This kind of environment – centralist, directive – is ostensibly one where the take-up of ICT in teaching and learning is likely to be feasible on a mass scale, simply because there is a high degree of (imposed?) congruence between national and local objectives.

However, it could be self-defeating, depending upon the cultural and social resistance to the changes proposed.

Technical competence will affect take-up in various ways. Web-based learning, for example, requires an infrastructure and a connectivity sufficiently well developed to make pervasive use by students and teachers a feasible and cost-effective proposition. In this context, the UK is at an advantage over even the USA, with the development of the JANET. Developing and underdeveloped countries are keenly aware of the need to improve their infrastructure accordingly. Technical competence is also required of the teachers and the students. Even in a relatively advanced environment such as that in the UK, there are still doubts about the level of ICT literacy among the HE community, and hence the many recommendations in reports such as that by Follett (1993) and the later work by Anstey (2000) about the need for developmental programmes.

Financial constraints are important. Major technological pushes require capital investment over a long period of time. This may require national-level initiatives, as for example in the case of HE (with the possible exception of the richest private universities) and an ability to fund them. The richer West is clearly at an advantage in this respect, and without external investment, developing and underdeveloped countries will lag behind for some time to come. It is still not clear how far ICT can be exported for teaching and learning purposes, although the Internet is beginning to provide a cost-effective way of doing so. It should be noted that the attraction of international students to western countries will have its own effect in the longer term as they take the skills and experience of ICT-based teaching and learning back to their own countries.

Summary

How, then, will ICT in teaching and learning best be developed? Senker (1994) suggests that: 'Japanese decision-making processes, based on consensus, are well adapted to manage the major decisions involved in investing in complex manufacturing systems.' Universities should therefore be well able to adapt to new technologies, given that they are at least ostensibly consensual organisations – at least at the macro level – where widespread consultation is valued.

However, there is more to it than that. As Green (1994) stresses:

> By being aware of and encouraging sensitivity to the way strategy communicates messages to organisation members, by framing these messages so as to strike a chord with what organisational members hold dear, and by recognizing strategy's potential for unifying disparate meanings across the organisation, the way to a more effective strategic management process is opened. But always the caveat remains: because meanings are located in peoples' minds, all attempts to manage them can, to use Eco's phrase, at best be tentative and hazardous acts of influence.

The application of ICT to teaching and learning is a complex issue. Strategic management techniques provide a valuable tool, but ultimately it is the people dimension that needs to be most carefully considered in any applications programme.

Bibliography

Anstey, P. (2000) *C & IT Skills: Developing Staff C & IT Capability in Higher Education*. Norwich: University of East Anglia.

Cooper, R.G. and Kleinschmidt, E.J. (1987) 'New products: what separates winners from losers?', *Journal of Production Innovation and Management*, 4, 169–84.

Crawford, C.M. (1991) *New Products Management*. Homewood, IL: Irwin.

Dearing, R. (1977) *Report of the National Committee of Enquiry into Higher Education*. London: Funding Councils.

Earl, M.J. (1989) *Management Strategies for Information Technology*. New York: Prentice Hall.

Follett, B. (1993) *Joint Funding Councils' Libraries Review Group Report*. London: Funding Councils.

Green, S. (1994) 'Strategy, organisational culture and symbolism', in E. Rhodes and D. Wield (eds), *Implementing New Technologies: Innovation and the Management of Technology*, 2nd edn. Oxford: Blackwell, 422–35.

Hackett, G.S. (1994) 'Investment in technology – the service sector sinkhole?', in E. Rhodes and D. Wield (eds), *Implementing New Technologies: Innovation and the Management of Technology*, 2nd edn. Oxford: Blackwell, 21–30.

Hayes, R.H. and Abernathy, W.J. (1994) 'Managing our way to economic decline', in E. Rhodes and D. Wield (eds), *Implementing New Technologies: Innovation and the Management of Technology*, 2nd edn. Oxford: Blackwell, 7–20.

Higher Education Funding Council for England (1997) *Information Technology Assisted Teaching and Learning in Higher Education*. London: HEFCE.

JISC (Joint Information Systems Committee) (1995) *Guidelines on the Production of an Information Strategy*. Bristol: JISC.

Schoemaker, P.J.H. (1993) 'Strategic decisions in organisations: rational and behavioural views', *Journal of Management Studies*, 30(1), 107–30.

Senker, P. (1994) 'Implications of CAM/CAD for management', in E. Rhodes and D. Wield (eds), *Implementing New Technologies: Innovation and the Management of Technology*, 2nd edn. Oxford: Blackwell, 338–48.

Twiss, B. and Goodridge, M. (1989) *Managing Technology for Competitive Advantage*. London: Pitman.

Case study 3

E-Lib and the EDDIS project

Introduction

I previously undertook an evaluative study of the EDDIS project (Baker, 2002). The main aim of this study was to identify the key drivers of project success or failure in order to improve the research and development 'rate of return' (Barras, 1990) in UKHE to general benefit. The results of the study are discussed under the heading 'evaluation' below. The study took the form of an iterative survey process (based to a certain extent on Delphi techniques) with a number of well-known ICT experts within and outside the UKHE LIS sector, some of whom had been involved in either the EDDIS or the e-Lib programme or both; others had only had a tangential contact with the programme and the EDDIS project.

Background

The Electronic Libraries (e-Lib) Programme (1995–2001) sought to change the way in which UKHE LIS delivered functionality, services and content to their users, with special emphasis on IT delivery. The investment was substantial (£60m) and stimulated over 100 projects (Tavistock Institute, 1998, 2000; ESYS Consulting, 2001). A significant number of IT development projects generated by e-Lib appeared to fail despite being well resourced and project managed using a recognised methodology (ESYS Consulting, 2001). This was particularly true of the higher profile, resource-intensive research and development projects such as EDDIS, Electronic Document Delivery – the Integrated Solution (EDDIS Consortium, 2000).

EDDIS

EDDIS was a major three-year project aimed at developing an integrated electronic bibliographic search, item locate, order, document delivery and account service for teachers and students in higher education. The project partners were: the Universities of Bath, Lancaster, Stirling and East Anglia (UEA), and the Bath Information and Data Service (BIDS). Fretwell-Downing Informatics (FDI) joined the project at a later stage. The British Library agreed to be an associate partner and to supply documents. UEA was the lead partner. At its simplest the system was to represent a process that transformed readers' requests into supplied documents (see Figure CS1.1 in Case study 1). The primary input of this system would be the requested document and the primary output the satisfied request. However, there were a number of secondary inputs and outputs (see Figure CS1.2 in Case study 1).

The bibliographic database would tell both the requestor and the library service of the existence and possible location of documents. The document warehouse could be local or remote; it might consist of several different warehouses. Once the request was satisfied, management information would be produced which helped management to monitor and modify services. For example, heavily requested documents initially supplied remotely could then be better supplied locally. Supplied documents could be stored in a local warehouse for general rather than individual usage. The ILDRMS (Inter-Lending and Document Request Management System) represented the transformation technology by which the request would be satisfied. In addition, the supplied documents could be the products of electronic publishing systems whereas the bibliographic data were typically created, manipulated and communicated automatically.

The project

The EDDIS consortium bid for almost £680,000 from the UK higher education sector's Electronic Libraries (e-Lib) programme for funds to develop this integrated approach. The project was due to conclude in 1997. The partners brought considerable expertise and know-how to the project. All had developed parts of the desired integrated system; four were working libraries keen to offer electronic services to their users. Between them, there was a considerable confidence that all the targets would be met. In summary, these were:

- integrated software covering all aspects of the request and delivery process (the EDDIS product);
- working services in the library sites covering at least 30,000 document requests per annum;
- interoperability with other systems;
- standards compliance (notably with American National Standards Institute (ANSI) standard Z39.50 for search and retrieve and International Standards Institute (ISO) ILL for the ILDRMS module);
- an exit strategy that would provide a cost-effective product for UKHE.

Change of direction and productisation aims

Initially, the consortium was to build its own software. However, the pre-existence of FDI's VDX software led to the company's addition to the consortium. The VDX software offered an opportunity to introduce the full cycle of search, retrieve, order, deliver and account management for both book and journal publications, whether in hard copy or electronic format. This was the basis of the EDDIS philosophy and the EDDIS project team entered into a partnership with FDI on the understanding that this would lead to a shorter time to market. Once a product was available, it was hoped that it would be possible to use Lancaster University's customer base for its existing ILLOS stand-alone ILDRMS product to sell the EDDIS version of the VDX product into UKHE. A 12-month extension was granted and funded by the e-Lib programme in order to allow for the development of a full business plan for the EDDIS roll-out phase. The extension project built on the earlier work of the EDDIS project. Two of the original partners only were involved in this extension phase: Lancaster University library and FDI. The extension project began with the aim of turning the results of the EDDIS project into a product that would be attractive to the UKHE sector. This proved not to be feasible.

Reasons for failure

A number of reasons can be put forward for the failure of the extension project to produce a product and, indeed, to realise a market.

The EDDIS project itself had not concluded with a working system that all partners felt was capable of being installed as a reference model nor as the basis of sales to university libraries.

The development of such a model had been the goal of both the original project and the extension. However, Lancaster staff never felt that they had something that was attractive enough or complete enough to market, other than in theory. In addition, there remained tensions between the new product and the existing Lancaster software. The ILLOS software was tried and tested and well liked by all who used it. Moving to FDI software represented a difficult unknown, especially in the context of incomplete and untested software.

The fundamental issues related to lack of a good-quality business case and the question of the attractiveness of the product. There is little doubt that the specification for the EDDIS system, if it had been implemented in full, would have led to a quality system. Indeed, people commented that this was a 'Rolls-Royce' system, but at Rolls-Royce prices. Was there a sufficient market for the product given comments like this? The fact that the ILLOS system was shared at relatively low cost with the UKHE community rather than sold at a true commercial rate was always going to make it difficult for an alternative approach to be accepted. This is not to underrate the amount of effort that went in to getting a viable business plan throughout the EDDIS project and the extension – a plan that would cover the commitments (e.g. Oracle and other licence costs), plus marketing and development costs.

The problem was exacerbated by the emergence of competing software at much lower prices (but much lower functionality). The market had moved on in the time since EDDIS had been conceived. FDI themselves had engaged with library system suppliers who were incorporating the software into their older products. The Lancaster software as it stood represented a stand-alone approach (although subsequent developments have increased the power and capacity of Lancaster's software). EDDIS could not easily be a stand-alone replacement to a stand-alone system.

Evaluation

The evaluative study found that few if any of the expected outcomes of the e-Lib programme (ESYS Consulting, 2001) were met to any great extent, although a number of benefits not listed as being outcomes from the original programme mission (Tavistock Institute, 1998) did emerge. These were all indirect results of the programme, which ended up concentrating on culture change rather than project and product delivery.

EDDIS was also rated as being below average in relation to other projects within e-Lib and compared with other IT projects outside the programme and the sector. It did not deliver any of its stated deliverables (EDDIS Consortium, 2000). Like e-Lib, however, it did deliver useful learning experiences and provided a base for future work, having documented the testing of technology. The project management itself was seen as being flawed; the same was true of risk management.

Although both e-Lib and EDDIS were visionary, there were serious operational and structural flaws. UKHE LIS was ill-prepared for a major technology development programme, as evinced by respondents' comments. EDDIS was criticised for its lack of clarity, over-ambition, lack of leadership, unstable partnerships and technical fragility. These comments fit in with the conception–reality gaps found in healthcare information systems (Heeks et al., 1999). Heeks et al. also draw attention to the kind of private–public sector gap (that is, between the LIS partners and FDI) in experience and understanding that was evident in EDDIS.

Summary

Software development is always a risk. e-Lib concentrated on the EDDIS project rather than others at least partly because everyone had high hopes for it; indeed, there was a sense that EDDIS could not be allowed to fail because the role it played was too central. A lesson for the future is perhaps not to mix high risk with centrality of role. In addition, a considerable amount of reliance was placed on one software company, although FDI always undercharged in a commercial sense and were perhaps led by other customers who were investing larger sums of money in VDX developments. However, there is much evidence to suggest that the value EDDIS and other e-Lib projects added to the VDX product was not lost to the UKHE community and projects such as Agora and Docusend (discussed in Case study 1) appeared subsequently to produce what EDDIS strove to achieve.

It was clear that many of the elements of good programme and project management were not used extensively or systematically. There needed to be a collective approach to the inculcation of technology management standards within the sector and guidance on partnerships with commercial organisations. A clearer definition of success would also have been valuable. This could have been articulated through descriptions of the extent to which R&D is meant to succeed by failure or the introduction of working services and commercially attractive products.

Bibliography

Baker D. (2002) *Why Technology Projects Fail*. MBA dissertation, Open University.

Barras, R. (1990) 'Interactive innovation in financial and business services', *Research Policy*, 19, 215–37.

EDDIS Consortium (2000) *EDDIS Extension Project: Final Report*. Bath, Lancaster, Norwich and Sheffield: EDDIS Consortium [the report includes the final report of the original EDDIS project].

ESYS Consulting (2001) *Summative Evaluation of Phase 3 of the eLib Initiative: Final Report Summary*. London: ESYS Consulting.

Heeks, R., Mundy, D. and Salazar, A. (1999) *Why Health Care Information Systems Succeed or Fail*. Manchester: University of Manchester Institute for Development Policy and Management.

Tavistock Institute (1998) *Electronic Libraries Programme: Synthesis of 1997 Project Annual Reports*. London: Tavistock Institute.

Tavistock Institute (2000) *1999 Synthesis of eLib Annual Reports: Phase 2 and Phase 3*. London: Tavistock Institute.

Case study 4

The Sudan

Introduction

I led a strategic technology management workshop in Khartoum to assist senior Sudanese LIS managers to develop a long-term plan for the country's future digital library development. The workshop employed a number of planning and forecasting techniques, as described in this book. The results of the exercise are described here.

Background

The Sudan is a low-technology country in terms of expected provision in a western country such as, say, Britain or the United States of America. The existing infrastructures of road, rail, electricity and water are relatively poor. Outside the capital, Khartoum, the standard of living is low, even by the benchmark of many African countries. There is a significant disparity between life and aspirations in the capital, where some 25 per cent of the population lives. There have been a number of regime changes in recent years and internal disputes and civil wars have not helped economic and social development. There is a relatively moderate

Islamic base to the society, educational provision and the government of the country.

There is nevertheless a significant push to develop a fast-follower status in terms of ICT applications. There are several reasons for this, and a number of advantages and opportunities that the country could readily exploit to significant advantage if the technology management strategy is properly formulated and implemented. Within this overall framework, the development of a substantial digital library presence is seen as strategically significant. The responsibility for delivering a national network infrastructure has been outsourced to a private company (Sudatel) and the Sudanese National Electronic Library (SNEL) has been set up. There is already a substantial network – provided by Sudatel – which offers Sudan the opportunity to provide the latest-generation information resources and services across the country, through use of high-quality cable and wireless networks, even to the remotest regions, using solar energy where more traditional means of providing power are not easily available.

But digital library systems are only as good as the content – and the organisation of information – that they carry. This requires library and information managers to take charge of the process of content creation. It is hoped and expected that LIS workers will now be able to play their rightful and professional part in the development of a national digital library framework that will bring Sudan up to a competitive level with the rest of the world in this new Internet age. There is certainly both the will and the capacity to achieve this over the next few years.

Digital library workshop

A five-day workshop on Digital Library Development was held at Sudatel headquarters in Khartoum for 20 LIS professionals in December 2002. The sessions and presentations covered all key aspects of digital libraries and awareness of needs and aims was significantly heightened during the course of the week. Feedback from the workshop was very positive, with a number of key action points and recommendations being identified and agreed. These were as follows:

- Setting up of a content task force and a technology focus group to oversee the strategic management and development of digital libraries in the Sudan.
- Development of a planning framework within which to implement programmes of projects encompassing all key aspects of digital library management.
- Instigation of a series of practical workshops covering specific aspects of digital libraries, including digitisation, licensing arrangements and metadata creation.
- Implementation of a series of pilot projects encompassing: licensing experiments, digitisation prototypes, metadata creation, interoperability demonstrators, MLE and VLE experiments, remote access trials, the development of embyro centres of excellence in data provision, digitisation and IT-assisted teaching and learning.
- High-level briefings with government ministers and other officials.
- Sharing of activities and results through the creation of websites, discussion lists, regular seminars, an annual review of the programmes of work and long-term partnerships with centres of excellence in other countries.

Aim and objectives of the workshop

The aim of the workshop was:

To consider the key issues relating to the management and development of digital libraries within a Sudanese context and to develop an action plan for participants, both individually and collectively.

This aim was agreed with the organisers and the participants before the commencement of the workshop. It was reconfirmed at the first session on day one. In order to achieve this aim, the workshop had the following main objectives:

- to identify the key strategic issues concerning digital library management;
- to consider key aspects of every strategic issue in digital library management;
- to review developments and good practice elsewhere;
- to reflect on current proposals and projects in Sudan;
- to develop additional, necessary expertise among the workshop participants.

By the end of the workshop, it was expected that participants would have:

- identified the key strategic issues relating to digital library development in Sudan;
- learnt about approaches to digital library management in the UK;
- developed action plans for applying best practice to the Sudanese environment.

The workshop leader covered the following topics – regarded as all the basic elements of managing and implementing a digital library – during the course of the five days:

- defining the digital library
- digitisation

- interoperability
- managed and virtual learning environments
- intellectual property right: UK practice
- JANET
- authentication
- information security
- strategic management of digital libraries.

These sessions were complemented by three presentations from Sudanese colleagues on: the Sudanese Digital Library development; building the Sudanese Digital Library; the Khartoum Academy of Technology. Access to a number of sites and services in the UK was arranged during the course of the week and participants were able to gain a good idea of digital library developments in the UK.

Next stage actions and recommendations

The workshop brought together a nucleus of senior librarians who have the capability and the capacity to lead the development of digital libraries in the Sudan. There was a strong wish to maintain contact in order to ensure that the enthusiasm and momentum generated by coming together for five days did not dissipate at the end of the workshop. A number of key actions were discussed and agreed by the participants during the course of the final plenary sessions. These are described below.

Strategic management

Given the early stage of digital library development in the Sudan, there is a need for strong, centralised strategic management that will ensure maximum sharing of good practice and common implementation of standards across the country. As part of this process, there needs to be a steering group looking both at interoperability of systems and services and at future technology development and application. As already noted, a digital library environment is only as good as the content accessed and created. There will also need to be a priority programme for negotiating access to high-quality content and for the digitisation of core materials within the Sudan.

Skills development

The workshop recognised that there is a significant need for a greater technical know-how among librarians at all levels. The workshop has raised awareness of the issues and the requirements; it is now necessary to ensure that practical skills are sufficiently available to manage and implement programmes of work effectively. The evaluation responses from the workshop have begun to highlight some of the areas where further training and skills development is required.

Planning framework

In order to ensure that the development of digital libraries is undertaken in a managed way, the workshop agreed that a planning framework needed to be put in place. The most important priority is to set up two planning groups: a content task force and a technical foresight group. These two groups will have a broad representation from within the Sudan, but also involve experts from other countries. As well as UK input, it was suggested that links should be made with Ethiopia, where similar initiatives are now planned. In addition, the workshop leader has strongly recommended to the workshop participants that representatives should be nominated to international standards bodies in order to ensure that Sudan is fully compliant with such standards and is also able to influence future developments. The first and most important area for such involvement is the Z39.50 implementation group (the ZIG).

Initial programmes of work

The two groups will be responsible for developing programmes of work involving a series of demonstrator or pilot projects and for ensuring widespread dissemination of the results to both the professional and the user communities. It is strongly recommended that such discussions should involve key potential suppliers and providers of content. The initial programmes of work should cover a three-year period, and form a series of joined-up projects looking at all aspects of digital library development. Ideally, the programmes of work should be overseen by an executive board and carried through by a programme manager. It will be essential that programmes of work are supported at the highest levels and that the groups are given the authority to act.

Pilot projects

The following pilot projects were agreed with the workshop participants:

- National site licensing experiments
- Digitisation prototypes
- Metadata creation experiments
- Interoperability demonstrators
- MLE and VLE experiments
- Remote access trials

- Embryo centres of excellence – the following were suggested: Sudanese Data Services; Sudanese Digitisation Service; Sudanese Centre for IT-Assisted Teaching and Learning. These centres all have UK parallels, with which the Sudanese equivalents could usefully be linked, with two-way exchange of information and expertise taking place.

High-level briefings

In this context, it was generally agreed that a high-level briefing of senior government and other officials should take place within 6–9 months of the workshop, with the two expert groups providing much of the input to the briefing sessions.

Content Task Force

The Content Task Force (CTF) should be responsible for:

- identifying available content within the Sudan;
- identifying possible content available from outside the Sudan and negotiating national site licensing arrangements;
- drawing up a priority list of Sudanese library and other collections to be digitised;
- drawing up a list of priority demonstrator projects within the field of content creation;
- taking responsibility for developing authentication and authorisation protocols that minimise the barriers to usage of the services provided.

Technology Foresight Group

The Technology Foresight Group (TFG) should be responsible for:

- assisting in the development of a robust infrastructure consonant with the development of latest-generation digital libraries;
- oversight of standards implementation within the Sudan. This will include: Z39.50, Dublin Core, Open Archives Initiative, Open-URL, E-Print, IMS, etc., and require representation on appropriate standards bodies;
- development of approaches appropriate to the Sudan in respect of emerging technologies. This could include WAP (wireless application protocol) and PDA (personal digital assistant) initiatives;
- providing a steer to the adoption of technologies and the use of third-party suppliers.

Sharing of results

It is crucial that the results of all activities, and awareness of developments, both nationally and internationally, are widely shared across the Sudan.

The following means of dissemination were agreed:

- creation of a digital library website;
- instigation of e-mail discussion lists among key LIS staff;
- regular seminars;
- an annual review of the programmes of work;
- long-term partnerships with centres of excellence in other countries.

A second workshop is planned for January 2004, at which the work since the last workshop will be evaluated and further actions discussed and decided.

Further reading

Aggarwal, R. and Rezaee, Z. (1996) 'Total quality management for bridging the expectations gap in systems development', *International Journal of Project Management*, 14(2), 115–20.

Albert, K.J. (1983) *How to Solve Business Problems*. New York: McGraw-Hill.

Ambler, S.W. (2001) 'Planning modern day software projects', *Computing Canada*, 27(4), 11–16.

Anstey, P. (2000) *C & IT Skills: Developing Staff C & IT Capability in Higher Education*. Norwich: University of East Anglia.

Artto, K.A. et al. (2001) 'Managing projects front-end: incorporating a strategic early view to project management with simulation', *International Journal of Project Management*, 19(5), 255–65.

Atkinson, R. (1999) 'Project management: cost, time and quality, two best guesses and a phenomenon; it's time to accept other criteria', *International Journal of Project Management*, 17(6), 337–42.

Baccarini, D. and Archer, R. (2001) 'The risk ranking of projects: a methodology', *International Journal of Project Management*, 19(3), 139–46.

Barras, R. (1990) 'Interactive innovation in financial and business services', *Research Policy*, 19, 215–37.

Belassi, W. and Tukel, O.A. (1996) 'A new framework for determining critical success/failure factors in projects', *International Journal of Project Management*, 14(3), 141–51.

Bradley, K. (1997) *Understanding PRINCE 2*. Bournemouth: SPOCE Project Management Ltd.

Bryde, D.J. (1997) 'Underpinning modern project management with TQM principles' *TQM Magazine*, 9(3), 231–8.

Chapman, C. (1997) 'Project risk analysis and management – PRAM the generic process', *International Journal of Project Management*, 15(5), 273–81.

Chatzoglou, P.D. and Macaulay, L.A. (1996) 'A review of existing models for project planning and estimation and the need for a new approach', *International Journal of Project Management*, 14(3), 173–83.

Chatzoglou, P.D. and Macaulay, L.A. (1997) 'The importance of human factors in planning the requirements capture stage of a project,' *International Journal of Project Management*, 15(1), 39–53.

Cicmil, S. (2000) 'Quality in project environments: a non-conventional agenda', *International Journal of Quality and Reliability Management*, 17(4), 554–70.

Clarke, A. (1999) 'A practical use of key success factors to improve the effectiveness of project management', *International Journal of Project Management*, 17(3), 139–45.

Cohen, L. and Manion, L. (1980) *Research Methods in Education*. London: Croom Helm.

Collins, T. and Bicknell, D. (1998) *Crash: Learning from the World's Worst Computer Disasters*. London: Simon & Schuster.

Davenport, T.H. (1994) 'Saving IT's soul: human-centered information management', *Harvard Business Review*, March–April, 119–31.

Davis, J. et al. (2001) 'Determining a project's probability of success', *Research Technology Management*, 44(3), 51–8.

Dearing, R. (1997) *Report of the National Committee of Enquiry into Higher Education*. London: Funding Councils.

Dey, P.K. (1999) 'Process re-engineering for effective implementation of projects', *International Journal of Project Management*, 17(3), 147–59.

Drummond, H. (1999) 'Are we any closer to the end? Escalation and the case of Taurus', *International Journal of Project Management*, 17(1), 11–16.

Dubernais, L. (2001) 'Yesterday's lessons, today's advanced tools, tomorrow's business success', *Buildings*, 95(6), 96.

EDDIS Consortium (2000) *EDDIS Extension Project: Final Report*. Bath, Lancaster, Norwich and Sheffield: EDDIS Consortium [the Report includes the final report of the original EDDIS project].

El-Sabaa, S. (2001) 'The skills and career path of an effective project manager', *International Journal of Project Management*, 19, 1–7.

ESYS Consulting (2001) *Summative Evaluation of Phase 3 of the Elib Initiative: Final Report Summary*. London: ESYS Consulting.

Feldman, J.I. (2001) 'Project recovery: saving troubled projects', *Information Strategy: The Executive's Journal*, 17(2), 6–12.

Follett, B. (1993) *Joint Funding Councils' Libraries Review Group Report*. London: Funding Councils.

Fowler, A. and Walsh, M. (1999) 'Conflicting perceptions of success in an information systems project' *International Journal of Project Management*, 17(1), 1–10.

Globerson, S. (1997) 'Discrepancies between customer expectations and product configuration', *International Journal of Project Management*, 15(4), 199–203.

Graham, J.H. (1996) 'Machiavellian project managers: do they perform better?', *International Journal of Project Management*, 14(2), 67–74.

Gray, R. (2001) 'Organisational climate and project success', *International Journal of Project Management*, 19(2), 103–10.

Hausschildt, J. (2000) 'Realistic criteria for project manager selection and development', *Project Management Journal*, 31(3), 23–33.

Heeks, R., Mundy, D. and Salazar, A. (1999) *Why Health Care Information Systems Succeed or Fail*. Manchester: University of Manchester Institute for Development Policy and Management.

Higher Education Funding Council for England (1997) *Information Technology Assisted Teaching and Learning in Higher Education*. London: HEFCE.

Hvam, L. and Have, U. (1998) 'Re-engineering the specification process', *Business Process Management Journal*, 4(1), 25–43.

Jaafari, A. (2000) 'Life-cycle project management: a proposed theoretical model for development and implementation of capital projects', *Project Management Journal*, 31(1), 44–53.

Jaafari, A. (2001) 'Management of risks, uncertainties and opportunities on projects: time for a fundamental shift', *International Journal of Project Management*, 19, 89–101.

Jelinek, M. and Schoonhoven, C.B. (1990) *The Innovation Marathon: Lessons from High Technology Firms*. Oxford: Blackwell.

Jiang, J.J. (2000) 'Project risk impact on software development team performance', *Project Management Journal*, 31(4), 19–27.

Johnson, J. et al. (2001) 'The criteria for success', *Software Magazine*, 21(1), 1–8.

Joint Information Systems Committee (1995) *Guidelines on the Production of an Information Strategy*. Bristol: JISC.

Joint Information Systems Committee (2001) *The Distributed National Electronic Resource*. Bristol: JISC.

Joint Information Systems Committee (2002) *Circular 1/02: Focus on Access to Institutional Resources Programme*. Bristol: JISC.

Kessler, E.H. (2000) 'Tightening the belt: methods for reducing development costs associated with new product development', *Journal of Engineering and Technology Management*, 17, 59–92.

Kharbanda, O.P. and Stallworthy, E.A. (1992) 'Lessons from project disasters', *Industrial Management and Data Systems*, 92(3).

Kingston, W. (2000) 'Antibiotics, invention and innovation', *Research Policy*, 29, 679–710.

Kirby, E.G. (1996) 'The importance of recognizing alternative perspectives: an analysis of a failed project', *International Journal of Project Management*, 14(4), 209–11.

Kloppenborg, T.J. and Petrick, J.A. (1999) 'Leadership in project life cycle and team character development', *Project Management Journal*, 30(2), 8–14.

Kumar, N. et al. (2000) 'From market driven to market driving', *European Management Journal*, 18(2), 129–42.

Kuprenas, J.A. (2000) 'Project manager workload – assessment of values and influences', *Project Management Journal*, 31(4), 44–52.

Laszlo, G.P. (1999) 'Project management: a quality management approach', *TQM Magazine*, 11(3), 157–60.

Lopes, M.D.S. and Flavell, R. (1998) 'Project appraisal – a framework to assess non-financial aspects of projects during the project life cycle', *International Journal of Project Management*, 16(4), 223–33.

McDonald, J. (2001) 'Why is software project management difficult? And what that implies for teaching software project management', *Computer Science Education*, 11(1), 55–71.

Munns, A.K. and Bjeirmi, B.F. (1996) 'The role of project management in achieving success', *International Journal of Project Management*, 14(2), 81–7.

Nellore, R. and Balachandra, R (2001) 'Factors influencing success in integrated product development (IPD) projects', *IEEE Transactions on Engineering Management*, 48(2), 164–75.

Noori, H. (1990) *Managing the Dynamics of New Technology: Issues in Manufacturing Management*. Englewood Cliffs, NJ: Prentice Hall.

Orwig, R.A. and Brennan, L.L. (2000) 'An integrated view of project and quality management for project-based organisations', *International Journal of Quality and Reliability Management*, 17(4), 351–63.

Pender, S. (2001) 'Managing incomplete knowledge: why risk management is not sufficient', *International Journal of Project Management*, 19, 79–87.

Pinto, J.K. (2000) 'Understanding the role of politics in successful project management', *International Journal of Project Management*, 18, 85–91.

Pitagorsky, G. (1998) 'The project manager/functional manager partnership', *Project Management Journal*, 29(4), 7–17.

Raz, T. and Michael, E. (2001) 'Use and benefits of tools for project risk management', *International Journal of Project Management*, 19, 9–17.

Rowley, J. (1993) *Computers for Libraries*. London: Library Association Publishing.

Sampath, R.S.V. (2001) 'PSM standards for effective project management', *Hydrocarbon Processing*, 80(5), 92-A.

Smith, P.G. and Reinertsen, D.G. (1991) *Developing Products in Half the Time*. New York: Van Nostrand Reinhold.

Stewart, W.E. (2001) 'Balanced scorecard for projects', *Project Management Journal*, 32(1), 38–54.

Sundbo, J. (1997) 'Management of innovation in services', *Service Industries Journal*, 17(3), 432–55.

Tavistock Institute (1998) *Electronic Libraries Programme: Synthesis of 1997 Project Annual Reports*. London: Tavistock Institute.

Tavistock Institute (2000) *1999 Synthesis of Elib Annual Reports: Phase 2 and Phase 3*. London: Tavistock Institute.

Thite, M. (2000) 'Leadership styles in information technology projects', *International Journal of Project Management*, 18, 235–41.

Thomas, S.R. (1999) 'Compass: an assessment tool for improving project team communications', *Project Management Journal*, 30(4), 15–25.

Thoms, P. and Pinto, J.K. (1999) 'Project leadership: a question of timing', *Project Management Journal*, 30(1), 19–27.

Tukel, O.I. and Rom, W.O. (2001) 'An empirical investigation of project evaluation criteria', *International Journal of Operations and Production Management*, 21(3), 400–16.

Twiss, B. (1992) *Managing Technological Innovation*. Harlow: Longman.

Twiss, B. and Goodridge, M. (1989) *Managing Technology for Competitive Advantage*. London: Pitman.

Uher, T.E. and Toakley, A.R. (1999) 'Risk management in the conceptual phase of a project', *International Journal of Project Management*, 17(3), 161–9.

Vadapalli, A. and Mone, M. (2000) 'Information technology project outcomes: user participation structures and the impact of organisation behaviour and human resource management issues', *Journal of Engineering Technology Management*, 17, 127–51.

Vandersluis, C. (2001) 'Almost all pilot projects lack measuring metrics', *Computing Canada*, 27(11), 13–18.

Van Gundy, A.B. (1988) *Techniques of Structured Problem Solving*. New York: Van Nostrand Reinhold.

Ward, S. (1999) 'Requirements for an effective project risk management process', *Project Management Journal*, 30(3), 37–44.

Wateridge, J. (1997) 'Training for IS/IT project managers: a way forward', *International Journal of Project Management*, 15(5), 283–88.

Webster, G. (1999) 'Project definition – the missing link', *Industrial and Commercial Training*, 31(6), 240–45.

Whetherly, J. (1998) *Achieving Change Through Training and Development*. London: Library Association Publishing.

Whittaker, B. (1999) 'What went wrong? Unsuccessful information technology projects', *Information Management and Computer Security*, 7(1), 23–30.

Williams, T.M. (1997) 'Empowerment vs risk management?', *International Journal of Project Management*, 15(4), 219–22.

Wirth, I. (1996) 'How generic and how industry-specific is the project management profession?', *International Journal of Project Management*, 14(1), 7–11.

Wright, J.N. (1997) 'Time and budget: the twin imperatives of a project sponsor', *International Journal of Project Management*, 15(3), 181–86.

Zwikael, O. et al. (2000) 'Evaluation of models for forecasting the final cost of a project', *Project Management Journal*, 31(1), 53–8.

Index

Printed in the United Kingdom
by Lightning Source UK Ltd.
101209UKS00001B/100-105